Laterite and
Landscape

Laterite and Landscape

M. J. McFARLANE

Formerly Department of Geography,
University of Nairobi,
Nairobi, Kenya

1976

ACADEMIC PRESS
London · New York · San Francisco
A Subsidiary of Harcourt Brace Jovanovich, Publishers

ACADEMIC PRESS INC. (LONDON) LTD
24—28 Oval Road,
London NW1

US Edition published by
ACADEMIC PRESS INC.
111 Fifth Avenue,
New York, New York 10003

Library of Congress Catalog Card Number: 76—1091

ISBN: 0—12—484450—2

Text set in 11/12 IBM Baskerville by Santype International Ltd., Salisbury, Wiltshire

Printed by Whitstable Litho Ltd., Whitstable, Kent

Introduction

Tropical geomorphology is a relatively new and rapidly expanding area of study. Slow to shed the stigma of its "accidental" nature as opposed to the "normality" of geomorphology in temperate regions, in recent years its unique characteristics have begun to be appreciated. Thomas' recent book on the subject has done much to enlarge the area of knowledge of the processes which operate to shape humid tropical landscapes. In particular he has given tropical weathering a status befitting its importance. Weathering is a fundamental geomorphic process which precedes all others in any environment, but in the tropics is paramount, and weathering processes and products play a major if not dominant role in modelling landscapes. Goudie's recent account of the weathering crusts, the end-products of weathering collectively termed duricrusts, assembles and extends existing data on crusts especially calcretes.

Laterite was first given a place in scientific literature nearly 170 years ago. There is now a very extensive literature to which reviews can hardly hope to do justice. There are remarkably few areas of agreement on how laterite developed and in particular on the precise nature of the environment conducive to its formation. Goudie has rightly pointed out that geomorphologists have in effect shirked the problem of crust development, partly because for various reasons they feel ill equipped to tackle it. He also, rightly I think, pointed out that it is highly desirable to understand crust development in relation to tropical landscape evolution since this is an essential prerequisite for the application of conservation to tropical environments. The present review is not therefore a general review of laterite literature, but a systematic discussion of laterite essentially for geomorphologists. Those in allied subjects, for example geology, may also find the approach useful. The main purpose has been to extract from a wide range of literature only that which appears to be relevant to the geomorphologist who wishes to use laterite as an index for landscape analysis. For this reason in many cases it discusses aspects which are peripheral to the themes of the various authors, who work in a variety of disciplines. It is considered that such licence is acceptable in an attempt to discuss only the subjects relevant to the relationship between laterite genesis and landscape development. Although it is by no means exhaustive even within these limits, the review has been fairly wide. Other students of laterite will inevitably use literature inaccessible to the present writer, but it is hoped that they may nevertheless find this discussion useful in their own assessment of the situation, since it summarizes the main concepts which have developed, and attempts to reconcile the many stark contradictions which appear in the literature. It also attempts to crystallize the main problems facing an understanding of laterite development. It aims to provide for new students of laterite a general guide into the complexities of the subject.

The review includes new data from Uganda, the home of what was reputedly the world's finest example of a laterite-capped planation surface, the Buganda Surface. This is an area that presents in striking form many of the problems encountered elsewhere in the tropics, and laterite specialists in other tropical areas may find in this recent work something of relevance to their own areas.

TERMINOLOGY

The laterite literature is heavily encumbered by problems of terminology. At present, one major difficulty is that different materials are often described by a single term, partly because of a wish to avoid overloading the large laterite vocabulary already existant. However, if a material cannot be described by the existing terms, for the sake of accurate description new terms must be used. This problem will be discussed more fully later (Chapter 8). In the mean time it should be borne in mind that the terms used by the different authors may not always relate to identical materials.

The terms used by the present writer are defined as follows:
(1) *Laterite.* This term is retained as a general name conforming with the loose usage of Sivarajasingham *et al.* (1962, p. 5). Thus it includes

"highly weathered material (1) rich in secondary forms of iron, aluminum or both; (2) poor in humus; (3) depleted of bases and combined silica; (4) with or without non-diagnostic substances such as quartz, limited amounts of weatherable primary minerals or silicate clays; and (5) either hard or subject to hardening upon exposure to alternate wetting and drying. The term used implies no restrictions other than those inherent in the properties defined, on the processes by which diagnostic properties have developed, or specific conditions of place or time as factors essential to such development. In this sense it includes Buchanan's laterite and hardened equivalents of it. In addition it includes certain highly weathered material in sesquioxide-rich humus-poor nodules that are hard or that harden upon exposure, though they may be surrounded by earthy material that does not harden, as well as masses of such nodules cemented together by sesquioxide-rich material".

(2) *Lateritic* is used as the adjective from laterite, and does not imply a lesser degree of development or a potential form.

(3) *Nodule* is used as a general term both in the sense defined by Bryan (1952) to include rounded lumps of a variety of compositions whether formed by "concretion" (i.e. accretion) or by centripetal enrichment, and also to include any rounded fragments of laterite inherited from a laterite crust (Sivarajasingham *et al.*, 1962).

(4) *Concretions and Mottles.** Concretions are well defined rounded bodies formed by centripetal enrichment. Mottles are formed by a similar process but do not have a well defined boundary. Like laterite, concretions and mottles may be hard or may harden on exposure.

(5) *Pisoliths and Ooliths.* Pisolith is used for well defined concretions sufficiently regular in size and form to be likened to peas or shot. It is limited to bodies of over 2mm diameter, as suggested by Twenhofel (in du Preez, 1954). Ooliths are similar but with a diameter of less than 2mm. Where the ooliths or pisoliths have a concentrically banded structure, they will be described as banded pisoliths or banded ooliths. Otherwise it may be assumed that the structure is not banded.

(6) *Carapace* is used to describe the lateritic crust found at or near the surface of many tropical soil profiles. It has no connotation of particular hardness; it may be exposed and hardened, or unexposed and soft but still affording a measure of protection to the underlying material. Maignien (1958) distinguished between "cuirasse" with its connotation of hardness and purely mechanical protection, in the

---

* It is unfortunate and confusing that the accepted term for a product of centripetal enrichment is a concretion while the concretionary process is defined as a process distinct from centripetal enrichment.

fashion of an armour plate, and "carapace" which may be softer. Since the degree of induration is usually determined by such arbitrary and subjective criteria as how hard one must hit it and with what, before it will break, such an ill defined separation of the terms seems to limit the value of the division, and it is rejected in this study in favour of the single term carapace which caters for both the indurated and unindurated crusts. The term "duricrust", introduced by Woolnough (1927) again implies hardness, which is difficult to define, and it is also tautological. It appears to offer no advantage over the term carapace and it is avoided in this review.

(7) *Planation surface* is used to describe a morphological feature, that is "An almost flat featureless plain showing little sympathy with structure, and controlled only by a close approach to base level. . . ." (Stamp 1961). It has no genetic connotation and as suggested by Brown (1961) it is used in preference to "erosion surface".

(8) *Erosion cycle* or *Cycle* is used much as defined by Davis in 1909 (in Stamp, 1961) to mean "The period of time during which an uplifted landmass undergoes its transformation by the processes of landsculpture, ending in a low featureless plain. . . .". It does not imply any regularity of the succeeding periods of time involved (OED), which in nature's cycles are rarely of regular length. Nor does it imply that the endpoint is identical to the point of initiation. The use of the term in many parts of the tropics seems more defensible than elsewhere, because periods of stability are well expressed here and the processes of weathering and erosion are more rapid. This has resulted in landforms more aptly described as surfaces than bevels. Hence erosion cycle is more apposite than erosion phase.

Acknowledgements

I am greatly indebted to the Department of Geology and Mines of Uganda for encouragement and support of my work in that country where I was first introduced to laterite: in particular for drilling cores and making others available for study, for providing access to their records, for preparing samples for DTA and for carrying out partial analyses. I would like to extend special thanks to the Commissioner, Dr C. E. Williams, and to Mr C. E. Tamale-Ssali who succeeded him; to Dr J. Almond, Dr P. Nixon, Dr P. Brock, Dr R. Mcdonald, Miss A. P. Lynam, Mr J. McKay, Mr J. B. Pollock, Mr H. C. Patel, Mr G. W. Sewanywa and Mr H. M. Simwogerere.

I was able to pursue further the study of the stratigraphy of Ugandan laterites through the kind help of Mr Bicewski of Uganda M.O.W. and Mr J. Alexander of Mowlems, Uganda, who drilled and blasted profiles for me.

I thank also Dr R. C. McKenzie of Craigiebuckler, Aberdeen for electronmicrographs and Dr A. Mehlich, Dr A. A. Theisen and Mr G. Hinga of National Agricultural Labs, Nairobi, for X-ray analyses of samples and for advice on the use of DTA equipment which they kindly made available to me. I acknowledge with thanks the use of field equipment lent by Uganda Museum and the use of air photos borrowed from the Department of Lands and Surveys, Uganda.

I am indebted to Dr M. Posnansky for identification of artefacts, to Dr J. A. Miller for advice on resistivity methods, to the late Professor J. Laruelle (R.U.Ghent) for making thin sections of laterite samples and to Mrs P. Anderson for preparing symaps.

I am grateful to Mr Ntege Lubwama, Mr Lule and Mr Kiwanuka, formerly of the Ministry of Education, Mengo, for effecting introductions to the owners of the land on which particular studies were made and I am pleased to record here my gratitude to the many kind and friendly people I met in rural areas of Uganda who tolerated my intrusion in their idyllic countryside.

I thank Professor S. J. Baker, Professor B. Langlands and Dr P. Temple for their kindness and encouragement when I was based at the Geography Department, Makerere University.

Many people familiar with other areas in the tropics very kindly found time to correspond with me about laterite, to answer questions and discuss points, thereby enabling me to place my Ugandan work in a wider context. In particular I thank Dr M. J. Mulcahy (CSIRO, Western Australia) and Dr L. T. Alexander (USDA) for their helpful contributions; also Professor B. C. King (Bedford College, London) and Dr J. V. Hepworth (IGS) who stimulated my interest in the Kyagwe area in which he formerly worked. I am also grateful to my husband who made it possible for me to see something of laterites in Kenya, Zaire and Republic of Central Africa.

My work in Uganda was financed by Science Research Council (NATO), the Goldsmith Company and the British Council, to whom I am extremely grateful. Photographic reproduction of diagrams for the manuscript of this review was financed by a grant from Nairobi University. The diagrams were prepared with considerable help from the drawing offices of the Geography Departments of University College, London and Nairobi University. The majority are the work of Mr G. d'Souza. Reproduction of the stereopairs (Plates 16 and 17) was by kind permission of Lands and Surveys, Uganda.

Finally, particular thanks are extended to Professor E. H. Brown (University College, London) who supervized my study and to Professor W. W. Bishop (Queen Mary College, London) who introduced me to Ugandan geomorphology and who has consistently helped and encouraged me throughout my research. To both Professor Brown and Professor Bishop I am grateful for encouragement to write this review. Its shortcomings are mine but any contributions it makes are in large part due to those here acknowledged.

17 Lunds Farm Road   M. J. McFARLANE
Woodley
Berkshire

Contents

CONTENTS

1

A Historical Review of Theories of Laterite Genesis

I.   1800—1910

A.   Laterite as a Rock

Until about the first quarter of the twentieth century laterite was regarded as a rock type and was discussed largely by geologists. The earliest descriptions, those of Newbold, described it as "a purplish, or brick-red rock" (1844), and a "singular rock, unknown in Europe" (1846b). In Australia the affinity of the material to what was then recognized as desert sandstone led to a delay in the adoption of the term laterite to describe comparable material (Darwin, 1844; Prescott and Pendleton, 1952), but in Africa and elsewhere, the term was more readily accepted for such "rock" (Bain, 1852).

Its status as a rock was reasserted about the first decade of this century, during the heated discussions about the relationship of laterite to bauxite (Maufe, 1933; Evans, 1910a,b). Crook (1909) stated that they were not to be confused, as bauxite is a mineral name and laterite is a rock name, and in 1910 he reasserted the importance of laterite as a petrographic term.

Even after the assumption of interest in laterite by the pedologists in the first half of the twentieth century (Scrivenor, 1933) descriptions of it as a rock still persisted. Possibly the fact that "it is usually shattered when struck a sharp blow with a hammer" (du Preez, 1949) is partly responsible for the persistence of this concept even in more modern writing (e.g. Johnson, 1959). Apart from the obvious property of hardness, the concept of it as a rock stemmed

1

largely from the now outdated division of the weathering profile into two distinct horizons (Reiche, 1950). In the upper parts processes due directly or indirectly to life, referred to as "soil forming", predominated. This was regarded as the domain of the pedologists. The lower parts of the profile were characterized by inorganic processes of weathering, the domain of the geologist. The occurrence of laterite at the surface was believed to be due to the removal of the "soil". It was believed to form at depth and was seen to belong more properly to the zone of inorganic alteration or weathering.

B. Laterite as a Weathered Rock — a Residuum

Apart from those few early workers who unreservedly believed laterite to be a rock in the strictest sense of being a deposit, most who described it as a rock qualified this by saying that it was a weathered or altered rock, or that the rock was the product of alteration or weathering of an original rock. Unlike other sediments which result from the removal of certain constituents from a parent rock and their subsequent deposition, this "sediment" was residual. It was what was left behind after other constituents had been removed (Babington, 1821; Benza, 1836; Clark, 1838).

How the differential removal of part of the rock occurred was not then understood. The formulation of the concept of deep weathering and differential solution under tropical conditions as the cause of the differential accumulation of the residuum is usually attributed to Russell (1889); but by 1821 Babington had already outlined the decay of hornblende and felspars into red oxides and porcelain earth from which the soft parts are washed away leaving a residuum. Shortly afterwards Newbold (1846b) characterized the formation of laterite as a "segregation and subsequent rearrangement of the different mineral particles in the substance of the rock itself, by a process in nature's laboratory, approaching to crystallization, better known than explained or understood". Russell (1889) provided some data on this process. He discussed the subaerial decay of rocks and the origin of the red colour of certain formations in the USA. He compared the greater solvent power of percolating waters in the tropics with that of cooler environments and suggested that the laterites of India, the red earths of Bermuda, and the terra rossa of southern Europe were all residua after the removal by chemical solution of considerable depths of the original rock.

The theory that laterite was the residuum of extreme weathering and differential removal of material by chemical solution became widely accepted. Glinka (1914) reinforced and elaborated Russell's (1889) theories of tropical weathering and formation of stable (laterite) residues. In 1932, the Imperial Bureau of Soil Science defined laterite as "a weathered rock product formed by the leaching of igneous and metamorphic rocks whereby the bases and much of the silica are removed leaving a residue containing alumina uncombined".

The concept of laterite as a residuum has persisted, often supported by convincing evidence (de Vletter, 1955; Hartman, 1955; du Bois and Jeffery, 1955; Hanlon, 1945; Van Bemmelen, 1941), but from about the first decade of the new century there was, generally, progressively less stress put upon the residual aspect and more upon the precipitationary aspect.

C. The Precipitationary Aspect of the Residuum

Although the stress laid on the residual nature of laterite would appear at first to imply that it is entirely a mechanical residue, clearly it was recognized even then that laterite is itself a precipitate. Two short-lived phases of mobility of the constituents were recognized. Prior to the accumulation of the residuum, the components became mobile and were precipitated in immobile or relatively immobile form. The mobility was short-lived, certainly less than that of the materials which were washed away, but a definite period of mobility was required for the minerals to regroup, after the breakdown of the parent materials. The *accumulation* of the precipitates was believed to be essentially a mechanical process and certainly the stress was primarily on the *immobility* of the precipitates to account for the accumulation of the residuum. The second phase of mobility was the subsequent rearrangement of the different minerals which were accumulated (Newbold, 1846b). Darwin (1844) for example described the formation process as "alluvial action on detritus abounding in iron". Increasingly towards the first decade of the twentieth century there are references to subsequent re-solution of the residuum and suggestions that the medium of re-solution and precipitation is groundwater e.g. Harrison and Reid (1910).

II.   1910—1960

Laterite as a Precipitate

The first suggestion that the role of groundwaters in the formation of laterite had been grossly underestimated was made by Maclaren

(1906), who stated that laterite is not the result of decomposition *in situ* of the rocks, but of the replacement of such decomposition products; the mechanism of the replacement being groundwaters. The groundwaters may have derived their mineral content from the underlying rock, but may also have brought it from sources far distant. With this concept, the stress on the immobility of the precipitates as the means of accumulation lost ground to the suggestion that the accumulations were in large part due to precipitation from groundwaters, and the source of the enrichment of these groundwaters was below the laterite but not necessarily local.

Campbell (1917) was the first to formulate definitely the concept of laterite as a precipitate. Since its precipitationary aspect was already well recognized, he must be understood to have meant that the accumulated minerals had a much greater degree of mobility than was formerly recognized, that is, that *the concentration was due to this mobility*. This was a major turning point in the development of theories of laterite genesis and was responsible for bringing the material into the sphere of interest of pedologists.

The way had already been paved by Harrison and Reid (1910) who, in their postulated stages of laterite formation, likened the leaching stage to the process of podsolization and stressed the most striking similarity of the topsoils in all profiles, regardless of parent material. Mennell (1909), supported by others, stated that laterite is independent of geology, implying that the material is more closely related to biospheric conditions than to parent material. Bishopp (1937) believed that the end-product was entirely dependent upon the attainment of a biochemical equilibrium. Subsequently the possible roles of abundant organic material and bacteria were frequently discussed, the climax of these lines of thought being the suggestion that termites were responsible for laterite formation (Erhardt, 1951).

Although there had been earlier tentative suggestions (Oldham, 1893), Glinka (1914) was one of the first to visualize laterites as soils, and Harrassowitz (1930) further popularized this idea. Vilensky (1925) devised a pictorial scheme representing the principal types of soils of the world and placed laterites in a thermogenetic division or torrid zone.

Campbell's suggestion (Campbell, 1917) that laterite is a precipitate was followed shortly by numerous studies of the many and varied ways in which iron and alumina could be mobilized and reprecipitated (e.g. Britton, 1925; Deb, 1949). These chemical studies by the pedologists appeared to reinforce the suggestion that laterite is a precipitate and more correctly described as either a soil or a soil horizon. By the 1930s, discussion of laterite genesis is found more commonly in pedological than geological journals (e.g. Hardy and Follett-Smith, 1931; Doyne and Watson, 1933). Whether it was regarded as a soil horizon or a soil, or merely an occurrence within a soil, its pedogenetic associations became firmly entrenched (Martin and Doyne, 1927, 1930; Beater, 1940; Humbert, 1948; Nye, 1954, 1955; Calton, 1959; Burridge and Ahn, 1965; Maud, 1965; Moss, 1965).

The concept of laterite as a soil, owing its concentration of minerals to a mobility attributable to biospheric conditions, coincided with the increasing popularity of a chemical rather than a physical definition of laterite. The adoption of chemical criteria of definition and the general adoption of laterite by the pedologists was not readily accepted by the geologists (Maufe, 1933; Pendleton and Sharasuvana, 1946). Scrivenor (1933), however, approved.

"Agronomists have adopted laterite and cradled it with their own offspring podzol. Unless we want to retain Buchanan's utilitarian definition, I suggest we accept the adoption. Buchanan's etymological child is now in excellent hands".

This adoption is significant to geomorphological studies. The "new" popular belief that laterite is a precipitate formed in the soil rather than a residuum of weathering had fundamental implications concerning the direction of movement of the laterite components. The older theories of downward leaching of soluble components and the residual accumulation of insoluble precipitates implies an overhead source for the concentrates. During the process of differential removal of material there is compaction of the original rock, so that the insolubles move either vertically downwards with this compaction, or downslope. The post-Campbell (1917) theories of precipitation in a soil from enriched groundwater implied the opposite direction of movement. This received support from the circumstantial evidence that pallid zones often underlie laterites (Maclaren, 1906; Simpson, 1912). Thus it came to be believed that enrichment is essentially upwards. The mechanisms most commonly evoked were a stable fluctuating water table and capillarity, and the necessary conditions were visualized as occurring in association with a planation surface in areas of marked seasons. This in turn appeared to agree with the association of laterite with grassland (Harrassowitz, 1930).

In short, the suggestion that laterite was a precipitate appeared to be in *general* agreement with the environment in which it was most commonly *seen* to occur, that is, the grass-covered planation surface with its stable but seasonally moving water table.

Much subsequent work dealt in detail with individual laterite samples taken from given profiles, or individual laterite profiles, or more rarely groups of profiles in catenary relationship. It is to this phase of laterite literature that Parry Reiche's comments are most applicable.

"Weathering is a complex subject. One of the outstanding difficulties in making its aquaintance through the literature has been the wealth of descriptive detail in most published accounts. This has tended to obscure both the principles involved and the objectives of the study" (Reiche, 1950, p. 6).

With the wealth of detailed information available, the relationship of laterite genesis to the environment is often obscured or overlooked. Furthermore, if this relationship is considered it appears more often than not that the new data does little to substantiate the post-Campbell concept of laterite as a precipitate. One or more elements of the "typical" laterite profile may be absent, and peculiarities in the mobilities of the concentrates are postulated to explain this. If the laterite occurs in a region atypical climatically or vegetationally, then climatic change is assumed to explain this. Where the laterite occurs on a surface other than a planation surface, further peculiarities are introduced to explain the accumulation. In short, most of the accounts involve explanations requiring mobilities *peculiar* to each particular situation. The wealth of data produced, largely by pedologists, leaves a geomorphologist with the unsatisfactory generalization that there are many kinds of laterite and many different environments in which it may form. In order to use laterite as a geomorphic criterion it becomes necessary to know which laterite pertains to which environment.

III.   1960 ONWARDS

Recent Compromises and Developments

The post-Campbell concept of laterite as a precipitate nevertheless survived. However, there has recently been some rethinking of this and earlier concepts, particularly by geomorphologists, in an effort to establish a relationship between laterite development and landscape development. The wealth of more precise data which has become available exposes the inadequacies of many of the basic assumptions upon which rested the theories of laterite as a precipitate. It has emerged that laterites are not entirely restricted to planation surfaces (a prerequisite of the "precipitation school"). Water tables do not fluctuate to the extent that had been assumed, and could not be responsible for the vertical translocation of material through hundreds of feet. Capillary action is also very limited. The apparently anomalous laterite profiles, those without either weathered, pallid or mottled zones, seem to be as common as the so-called typical profiles. The distribution of laterite requires something more than climatic change to explain the numerous apparently anomalous occurrences. The broadly synchronous upward and downward movements of solutions to explain both the desilicification of the whole profile and also the upward enrichment of the laterite from the pallid zone, can find little support from the detailed chemical studies made. In the face of this confusing array of contradictory evidence pedologists have tended to maintain their original stance, believing laterite to be essentially a precipitate. With the development of more refined techniques attention has been focussed more and more upon the intricacies of the reactions believed to have occurred during the development of the many and varied types of laterite. Such studies are generally unrelated to the environment of the laterite. Geologists and geomorphologists, on the other hand, without denying that there may be variations in the mode of laterite genesis are more concerned with establishing the significance of those basic generalizations which cannot be ignored. Laterites are unique to the tropics and they are associated with low relief. There must therefore be some fundamental similarity in the genesis of these various laterites that hinges on the tropical environment and the condition of low relief.

Geomorphologists may be criticized for their lack of concern with the physicochemical intricacies of laterite genesis. However, it seems pointless to propound theories about the means of mobilization of iron in a profile when simple arithmetic shows that even if all the iron in a weathering profile can be mobilized the depletion of the pallid zone cannot adequately account for the quantity of iron accumulated in the laterite. The source of the enrichment is initially more important than its precise means of translocation. Likewise, it seems irrelevant to strive for an understanding of the chemistry of iron mobility in association with a fluctuating water table when it is abundantly clear that even if iron could be mobilized, transported and redeposited by a rising and lowering water table, no water table exists which moves seasonally through some 200 ft (61 m), which is the scale of some "typical" laterite profiles.

In their efforts to explain laterite genesis from the evidence offered by its broader environment, geomorphologists have found promise in some form of synthesis between the old model of down-wasting with residual accumulation, and the later model of precipitation in a planation surface profile. These attempts at synthesis are dealt with more fully elsewhere (Ch. 10). They are notable in that they reject one or more of the basic dogmas upon which rest theories of laterite as a precipitate. Thus, du Bois and Jeffery (1955) attempted to explain laterite formation in Uganda without the synchronous development of a pallid zone. The circumstantial evidence of the common coexistence of laterite and pallid zone is not accepted by them as indicating that the development of these two horizons involves exchange between them. Trendall (1962), in an even more iconoclastic attempt to look afresh at laterite genesis and landscape development suggested that laterite does not develop on planation surfaces at all, but that each laterite-capped mesa develops independently on an interfluve and is essentially acyclic. Unlike du Bois and Jeffery (1955), Trendall did, however, incorporate pallid zone development in his theory of laterite genesis. More recently the association of laterite development with planation surfaces in Uganda has been reasserted (McFarlane, 1971) and doubt thrown on the originality of the existing high relief of the laterite mesas in that area. While recognizing the precipitationary aspect of the laterites, this work suggests that, as the earlier workers believed, the *accumulation* is essentially residual, and in support of du Bois and Jeffery (1955), that the pallid zone is not a contemporaneous development.

It is interesting that the need for a synthesis of ideas has been felt most strongly by those working in Uganda, the home of such a reputedly classical example of a laterite-capped planation surface. Certainly the present situation is one in which there is much scope for very fundamental rethinking about laterite genesis. The history of the study of laterite genesis suggests that it is from a more detailed study of the environment of laterite that advances may be made in the understanding of its development.

2

What is Laterite?

To define laterite is extremely difficult. The earliest descriptive definitions failed and were replaced by ones based on the chemical contents or ratios of these components. These also proved inadequate and the more precise pedological concept of laterite as a soil type superceded them, but left the originally described azonal material effectively nameless. The evolution of the various concepts of laterite clarifies some of the problems of laterite genesis as a whole and strongly suggests that the name laterite be retained in the sense of an azonal occurrence of varied morphology related to the different geomorphic environments of formation. A soil classification which avoids the term laterite for zonal soils, retaining it only to describe the azonal occurrence, is desirable to alleviate the current conflict between pedological and geomorphological concepts of laterite. A geomorphic classification of the azonal occurrences would greatly facilitate their use in landscape analysis.

I. PHYSICAL DEFINITIONS

In the nineteenth century laterite was defined in terms of its physical properties, hardness and colour.

A. Hardness

Buchanan (1807) originally described laterite as material which was initially soft enough to be cut into blocks by an iron instrument, but which became hard as brick on exposure to air (Buchanan, 1807, p. 440). It was for this reason that it was suitable for making bricks;

9

and it is this utilitarian aspect which Buchanan chose to stress in giving the name laterite (*later*: Latin, brick). Subsequently Buchanan found material in Malabar which was hard while still in the ground and not exposed to the air. He was apparently puzzled by this but recognizing its relationship to his original laterite he called it brickstone. Later he interchanged the words brickstone and laterite, since it was obviously impossible for him to decide whether or not indurated material found on the surface was formerly soft. Thus, although the ability to harden on exposure was initially stressed and this emphasis has persisted to a limited extent (Fermor, 1911; McNiel, 1964; Marbut and Manifold, 1926) there was a very rapid extension of the term to include material which was hard whether or not the hardness was original.

The value of *ability to harden* as a suitable definitive characteristic was doubtful for other reasons. Blanford (1859) had noted that some of the material underlying the laterite "hardened" on exposure, and Harrison and Reid (1910) described certain mottled material which also did this. Since any clay will harden to some extent on exposure, the precise meaning of "hard" was questioned, and definition of the hardness peculiar to laterite was hampered by imprecise culinary adjectives, for example cheesy or doughy (Kellog, 1962). Russell (1962, p. 560) described the *unindurated* laterite as being "as soft as cheese", while Doyne and Watson (1933) noted that the mottled zone under the laterite "tends to harden *into* a cheesy consistency on the exposed surface" (present writer's italics).

The property of absolute hardness rather than ability to harden rapidly gained popularity as a descriptive criterion (Newbold, 1844; Martin and Doyne, 1927; Doyne and Watson, 1933). Pendleton and Sharasuvana (1946) defined laterite as "an illuvial horizon . . . with slag-like cellular or pisolithic structure, and of *such a degree of hardness that it may be quarried out* and used for building construction" and concluded their description with the comment that "most laterite is usually shatterable when struck a sharp blow with a *sledgehammer* . . ." (present writer's italics).

Many authors compromised by including both the ability to harden and the property of hardness as definitive characteristics of laterite (e.g. Kellog, 1949). The US Soil Conservation Service definition (1960) of laterite (under the Greek guise of "plinthite") notes that the original formation may be hard.

In short, although initially defined as having the ability to harden, the difficulty of determining the originality of the hardness of many laterites, coupled with an inadequate measure of the hardness of the product, led to the use of the term laterite to describe a variety of hard and potentially hard materials.

B. Colour

The range of material described by the name laterite was further extended by efforts to use colour as a definitive characteristic. Buchanan (1807, p. 440) noted that laterite contains a very large quantity of iron in the form of red and yellow ochres. Newbold (1844) then described it as a "purplish or *brick-red* porous rock" (present writer's italics). Here the seeds were sown for the belief that laterite was so called because it is red, as are bricks. However, Walther (1916) is cited (Prescott and Pendleton, 1952) as the instigator of the mistaken assumption that Buchanan named it laterite because of its colour. Believing the red colour to be a more consistent and meaningful criterion than hardness, Walther suggested that the word should be used for all red-coloured alluvia. The subsequent adoption of the term by pedologists and its introduction to cover several sorts of zonal and intrazonal soils, furthered this indiscriminate use of the term; for there "is a gradual increase in the proportions of warm (especially red) hues in zonal soils . . . from microthermal to tropical climates" (Thorpe and Baldwin, 1940). The result was that it became customary for travellers in the tropics to refer to anything red as laterite (Martin and Doyne, 1927), and by the beginning of the century the term was rendered almost completely meaningless.

II.   CHEMICAL DEFINITIONS

The turn of the century saw attempts to put laterite on a proper scientific basis and to define it by its chemical content.

A. The Iron Content

The assumption that iron was responsible for the characteristic hardening and should therefore be regarded as the critical ingredient was not without chemical evidence. Buchanan (1807, p. 400) originally implied this, and subsequently many iron-rich laterites were described (e.g. Simpson, 1912; Pendleton and Sharasuvana, 1946; Pendleton, 1941; Mulcahy, 1960; Moss, 1965; Humbert, 1948). Maignien (1958) treated the terms lateritization and ferrallitization as synonymous. However, it was not entirely clear that laterite owed its characteristic hardening ability to an iron content that was high in absolute terms. Iron was often shown to be a cause of induration (Marbut and Manifold, 1926; Martin and Doyne, 1927; etc.). Van der Merwe and Heystek (1952) for example described a variety of laterites which they defined in terms of breakdown of kaolinite to free alumina rather than the iron content. However, none of these hardened except those which they qualified with the term ferruginous. Iron enrichment and hardening seem thus to be associated in some way. Nevertheless such descriptions often related to concretion formation rather than to the massive laterite which is initially soft but hardens on exposure like Buchanan's laterite. The hard iron-rich pisolithic, oolithic or concretionary materials described are often surrounded by clay and so will never harden *en masse*. Only if the clay is removed so that these concretions are left in juxtaposition, where they may be subsequently bonded together, do they form massive laterite, and it was doubted if such re-sorted concretionary material qualified for the name laterite (Evans, 1910a; Crook, 1909). Moreover there are numerous materials unrelated or only loosely related to laterite which may be cemented by an iron matrix (Martin and Doyne, 1927; E. M. Driscoll, personal communication). Clearly the absolute iron content alone of a material is of limited value as a diagnostic criterion for the definition of laterite if it is insufficient to differentiate between these various materials.

Nor is the significance of absolute iron content in Buchanan's laterite very clear. Pendleton and Sharasuvana (1946) gave analyses of materials in Siam extensively used for building. Their data would suggest that even the "clay" (if we disregard the hard concretions) owes its hardening ability to a high iron content because out of the 35 samples analysed only three had more Al_2O_3 than Fe_2O_3 and there was usually two or three times as much of the latter as the former. However, Prescott and Pendleton (1952, p. 23) gave the chemical analyses of what they described as good building material from a wider variety of sources, showing that not only iron-rich (40.8% Fe_2O_3), but also relatively poor material (4.4% Fe_2O_3) is suitable. The ability of the "clay" to harden *en masse* with such a low iron content is interesting in the light of observations (Fripait and Gastuche, 1952) that for a natural kaolinitic clay 30% iron oxide is needed to saturate the clay surfaces before microconcretion formation begins. Presumably then, the formation of clays which are soft but which will harden on exposure is not to be entirely likened to the formation of hard concretions. This suggests that it is something more than the absolute iron content which is responsible for the characteristic ability to harden *en masse*. Harrison and Reid (1910) warned that the ability to harden would not be found in the chemical composition of laterites. They considered this to be due to the nature of the changes on dehydration of the material. Scrivenor (1910a) also attributed hardening to the redistribution and partial dehydration of the ferric hydrate.

The suggestion is clearly that the absolute iron content of a "clay" is no indication of whether or not it will harden, as does Buchanan's laterite. The iron content must be in a suitable *condition*, presumably with the ability to dehydrate, and it must be suitably distributed within the material. To attempt to define laterite in terms of the behaviour and distribution of the iron is at present impossible. All that can be said is that if iron is of sufficient quantity, of suitable condition and suitably arranged within the clay so that the whole will alter irreversibly to a hardened mass when the internal moisture regime is altered by exposure to more extreme conditions of wetting and drying, then the clay is a laterite. This is no more precise, scientific or practicable than the original definitions based on its physical properties.

B. The Alumina Content

Further attempts to define laterite scientifically concerned the alumina content. This occurs predominantly in two forms; the silicates (i.e. the kaolins, loosely termed "clay" by the early workers) and the hydrates or hydrated oxides. The hydrates can be regarded as the result of desilicification of the alumino-silicates. In other words, if one considers desilicification as synonymous with lateritization (Hardy and Follett-Smith, 1931; Martin and Doyne, 1930; Humbert, 1948; Van der Merwe and Heystek, 1952 etc.) then the hydrates are the end-product, while the kaolins are still capable of further desilicification.

The pre-twentieth century workers almost without exception saw lateritization as the end-product of desilicification of rocks by subaerial weathering; and they presumed it to be a mixture of clay (kaolins) and hydrates of iron. The earliest analyses demonstrated the presence of hydrates of alumina (Bauer, 1898; Holland, 1903; Warth and Warth, 1903). Hydrated alumina was already known as bauxite from the type of area of Baux in France, and without detailed knowledge of the nature of either the European or the tropical materials, it was not altogether an unnatural step to liken them. Thus it was considered by many (e.g. Holland, 1903; Warth and Warth, 1903) that "Except for the impurities present, laterite is identical with the bauxites of the northern hemisphere" (Martin and Doyne, 1927). The *indicator* of the end-point of desilicification or lateritization was the presence of free alumina; and for many workers free alumina became the essential diagnostic constituent (Martin and Doyne, 1927, 1930; Crook, 1909; Evans, 1910a,b; etc.). Iron was considered to be a major but non-diagnostic impurity, occurring in variable amounts (Doyne and Watson, 1933; Martin and Doyne, 1930; Holland, 1903). Its presence did not indicate that the end-product of the process of lateritization (desilicification) had been reached. Thus, the laterites which Maclaren (1906) described were iron-rich, but nevertheless he stated that any theory of laterite formation must take into account the absence of kaolin.

There was considerable dispute as to whether the name laterite should be discarded in favour of the name bauxite (Crook, 1909, 1910; Evans, 1910a,b; Scrivenor, 1910a,b,c, 1932). Cutting the Gordian knot of the controversy it seems that those who regarded free alumina as a diagnostic feature of what had been called laterite were divided between:

(a) those who favoured the stretching of the term laterite (as originally defined with the ability to set and be used for brick-making) to include material which may or may not be iron-rich, but is characterized by extreme desilicification and the presence of free alumina, and

(b) those who favoured the restriction of the term laterite to Buchanan's type of building material, and the stretching of the term bauxite to describe laterites which were characterized by free alumina.

This discussion reached a peak about 1910, but was entirely premature because it was not then known whether Buchanan's laterite was in fact characterized by the hydrate or the silicate! Pleas were made for the Indian geologists to put an end to the controversy by analysing it (Maufe, 1933; Scrivenor, 1932). One-hundred and thirty years after laterite was first described by Buchanan and some 30 years after the "is laterite bauxite?" controversy began, Fox was sent to collect samples of Buchanan's laterite for analyses (Fox, 1933). It was found to vary from "limonitic haematite to argillaceous or siliceous limonite" (Scrivenor, 1933, 1937). Laterite as originally defined was characterized by the aluminosilicate rather than the hydrate of alumina and was chemically quite different from the material which inherited its name. The case seemed strong for the use of another term to describe the alumina-rich material, such as bauxite, as Scrivenor suggested.

Out of this conflict certain facts emerged quite clearly. There was confusion of two materials.

1. SILICATES of alumina   a. with iron enrichment
                                (Buchanan's laterite)
                            b. without iron enrichment
2. HYDRATES of alumina   a. with iron enrichment
   (Bauer's laterite)   b. without iron enrichment

Buchanan's laterite happened to be 1a, that is, clay with iron enrichment. Scrivenor (1937) cited Fermor's argument that it was merely fortuitous that Buchanan did not encounter the material on the Deccan before the material he first described. Had he done this, it might have been the more aluminous varieties which would have been the type material.

It would seem from this that there was a choice. No one wished to call 1b laterite (Scrivenor, 1910c). So either laterite is defined in terms of iron enrichment, or it is defined in terms of desilicification, regardless of iron content. The difficulties surrounding definitions based on iron enrichment have already been outlined. Apart from the fact that they exclude the type material (!) the problems of definitions based on desilicification are perhaps equally great. For example, the synonymity of lateritization and desilicification assumes that materials characterized by kaolin could be further desilicified, that is, kaolin indicates an incomplete process of lateritization. However, the end-product of desilicification or lateritization need not be the hydrates in all circumstances. It may differ with lithology (Harrison and Reid, 1910; Scrivenor, 1937) and biospheric conditions (Bishopp, 1937). Other factors may also inhibit desilicification (Grim, 1953). Thus, the definition of laterite as the *end-product* of the process of desilicification (the main argument of those favouring the presence of the hydrate as diagnostic) is confused by a lack of understanding as to what the end-product is under varying environmental conditions, so that the term could not be used genetically.

A second more practical problem is that there does not appear to be any clear separation between clays and hydrates, the two usually occurring together in varying proportions (Holmes, 1914). Clays and the hydrates may occur inextricably intermixed (Harrison, in Bishopp, 1937), and bauxite may occur as segregations in the laterite (iron-rich clay) (e.g. Dey, 1942). Other workers found the more alumina-rich material to underlie the iron-rich within the profile (Mattson, 1941; Fox, 1927; Lacroix, 1913), and some found the reverse (Frasché, 1941; Van Bemmelen, 1941; Fisher, 1958). The separation of lateritic and non-lateritic materials is thus once again subjective and unscientific.

C. Iron and Alumina

The difficulties of defining laterite by iron or alumina content alone led to the inclusion of both as diagnostic criteria. (This would include Buchanan's laterite, 1a, and Bauer's 2a and 2b, but would exclude 1b.) Thus laterite came to be regarded as a residuum of the process of desilicification, rich in alumina, iron or both (Humbert, 1948; Fermor, 1911; Simmons, 1929; King, 1962; Frasché, 1941; etc.).

With such a broadening of the term it becomes necessary to introduce some qualification, to differentiate between the various types. Scrivenor (1937) suggested that laterite could be divided into siliceous, ferruginous or aluminous varieties, while highly aluminous laterite could be called bauxite. De Chételat (1938) proposed that a laterite should be regarded as aluminous or bauxitic if it contains not less than 50% aluminium and 20% Fe_2O_3. He regarded normal laterite as having equal proportions of iron and aluminium, and "cuirasses éminement ferrugineuses" as having 55% Fe_2O_3. More recently, Dury (1969, p. 80) suggested the following classification based on the essential chemistry of the duricrusts.

Silitic	SiO_2
Siallitic	SiO_2, Al_2O_3
Fersilitic	Fe_2O_3, SiO_2
Fersiallitic	Fe_2O_3, $FeOOH$, SiO_2, $Al_2O_3 nH_2O$, $AlOOH$
Ferrallitic	Fe_2O_3, $FeOOH$, $Al_2O_3 nH_2O$, $AlOOH$
Ferritic	Fe_2O_3, $FeOOH$
Fermagnitic	Fe_2O_3, MnO_2
Tiallitic	TiO_2, $Al_2O_3 nH_2O$
Allitic	$Al_2O_3 nH_2O$, $AlOOH$

Such classifications have their applications, but, as Fermor (1911) pointed out, the separation of the different types is arbitrary. Although based on chemical content, the definitions of these different kinds of laterite therefore lack scientific precision. For example, an aluminous laterite is a bauxite when it is an economic source of alumina (Van Bemmelen, 1941; Hanlon, 1945). With improved extraction techniques (Gozan and Vera, 1962) what may have been an aluminous laterite yesterday might become a bauxite tomorrow. What is more important is, as Fermor (1911) also pointed out, the lack of a consistent relationship between the physical properties of the materials and their chemical content. This is clearly demonstrated by studies in Uganda (McFarlane, 1969) where quite distinct physical types of laterite cannot be distinguished by their chemical content. It is the form in which their content occurs, that is the clay mineralogy, which varies consistently in accordance with the physical properties of the laterites. Classification of laterite in the field cannot therefore be based on chemical content, nor would this appear as a desirable criterion if it groups together some physically different laterites and separates others that are physically alike (McFarlane, in press).

In short, although scientific procedures have allowed the chemical constituents of laterites to be identified and their quantity measured, definitions of laterite in these terms have been unsatisfactory and are often no more precise than the older definitions they were designed to replace. The physical properties of laterite cannot be circumscribed in terms of iron and/or alumina content (Sivarajasingham *et al.*, 1962, p. 8). As investigation techniques improve, it may become possible to define the part that this content plays, but it can hardly be of practical value for the recognition of laterite in the field.

D. SiO_2/Al_2O_3, SiO_2/Fe_2O_3, $SiO_2/Al_2O_3 + Fe_2O_3$ and Other Ratios

Other attempts were made to define laterites in terms of silica sesquioxide ratios. Lateritization or desilicification results in a progressive change in the proportions of silica and sesquioxides present in the weathering material, that is in the silica sesquioxide ratio (Harrison and Reid, 1910). Iron enrichment, alumina enrichment and desilicification can therefore be expressed by these ratios, which lower as the silica content of a clay is reduced and iron and alumina are accumulated.

Those who regarded the breakdown of kaolinite to free alumina as a measure of lateritization stressed the silica/alumina ratio (Van der Merwe and Heystek, 1952; Martin and Doyne, 1927; Crowther, 1930). Thus since the silica/alumina ratio of kaolin is theoretically

17

1.18, where a ratio lower than this occurs, there is assumed to be free alumina (Hardy and Follett-Smith, 1931). However, this figure is rarely used as a measure of lateritization, for in practice there is evidence that appreciable quantities of free alumina may occur even when the ratio is higher. For example, Sen (Sen *et al.*, 1941) found free alumina and iron oxide in considerable quantities in soils with a ratio in the clay fraction of over 2.0 and as high as 2.88 in one case. Martin and Doyne (1927), who introduced the silica alumina ratio as diagnostic of laterite, initially suggested that a clay with a ratio of less than 1.35 was to be considered laterite, and with a ratio of 1.35–2.0 it was lateritic. Later (Martin and Doyne, 1930) they modified this opinion, suggesting that 2.2 or 2.3 might be the upper limit of laterite in certain circumstances. Moorman and Panabokke (1961) used the ratio 2.0 to divide lateritic and non-lateritic materials.

The reasons for the variety in choice of the critical ratio is beyond the scope of this work, but it is clear that the use of silica alumina ratios as a means of defining laterite, or even determining the presence of free alumina is more arbitrary than would at first sight appear. Van der Merwe and Heystek (1952), although using breakdown of kaolin to alumina as a means of measuring and comparing the *degree* of lateritization of materials, indicated that this ratio was at best an imprecise measure of the process. They noted that

> "when the ratio is less than 2.0 as in colloids of laterite soils, it is fairly safe to postulate that the main minerals present are those of the kaolin group, gibbsite and iron oxide".

Certainly as a means of defining laterite, with a view to inter-regional comparison, the ratio is often misleading. Thus, laterite concretions have been described with a silica alumina ratio of 4.2 (Doyne and Watson, 1933), and a podzol described by Barshad and Rojas-Cruz (1950) had an ironstone layer with a ratio of 2.37.

Some of the difficulty is caused by the use of the whole soil rather than the clay fraction for analyses (Martin and Doyne, 1927). Free silica occurs as inherited quartz in many laterites (Sivarajasingham *et al.*, 1962, p. 10). Anaylses of the whole soil therefore give disproportionately high ratios. However the limiting of such analyses to the clay fraction does not entirely remedy this situation, as there is evidence of the occurrence of free inherited quartz even in this fraction (Mohr and Van Baren, 1954, p. 379). The ratio can be misleading whatever fraction is used, *unless it is accompanied by determination of the free quartz as well as total silica* (Pendleton and Sharasuvana, 1946).

Silica sesquioxide ratios are certainly difficult to determine in a form which makes interregional comparisons valid and, as different methods have been used, the value of comparison of the results is dubious (Pendleton and Sharasuvana, 1946). Both total and molar ratios have been used. Crowther (1930) discussed the merits of these and concluded that they vary with circumstances, there being little to choose between them.

If the method of determination is defined and if it is applied to comparisons of fractions within a soil (Doyne and Watson, 1933) or to different horizons of a soil (Barshad and Rojas-Cruz, 1950), this is often a useful means of description and comparison. As a means of comparing laterites in general it is of little value because of the lack of standardization of techniques. As an absolute means of defining laterites it is often quite meaningless (Pendleton and Sharasuvana, 1946). It may define as laterite material which is not hard or have the ability to harden (Van der Merwe and Heystek, 1952; Moorman and Panabokke, 1961) and may even exclude material which does harden.

E. The Pedological Concept of Laterite

Definitions of laterite progressed from those based upon subjectively assessed physical properties, to equally unscientific ones based on arbitrarily defined quantities of the included materials or ratios of these.

The pedologists' opinion, summarized by Thorpe and Baldwin (1940), seems to have been that Buchanan's "laterite" was not a laterite by their definition. Their justification of this remarkable reversal of the situation was that the name laterite was stretched at an early date to include both red podzolics and red laterite soils. The latter are the residuals of tropical weathering (desilicification) comparable to what d'Hoore (1954) later described as "relative accumulations" and for which no name then existed. To call them laterite soils was not unreasonable. The former, the red podzolics, have a marked zone of iron accumulation (precipitation) in their lower horizons. We are told (Thorpe and Baldwin, 1940) that pedologists had worked out a general hypothesis of the pattern of soil formation in the tropics, to which Buchanan's laterite does not conform chemically, and the reason for this is that he was in fact

19

describing the lower horizons of a podzol from which the upper horizons had been eroded. At the time when the name laterite was "adopted" by pedologists, the name podzol was already established. It seemed undesirable to use the word laterite for both the residual soils and the lower horizons of a podzol, and the name was already widely used to describe the residual soils. Apparently (Thorpe and Baldwin, 1940) it was the recognition of this confused state of affairs that caused Marbut to draw a hasty distinction between the laterite soils of the pedologists and "groundwater laterite" (Buchanan's lower horizon podzol). To the pedologists the terminological situation appears to be:

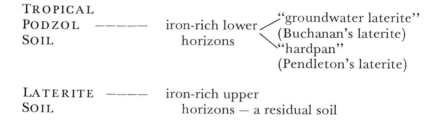

TROPICAL
PODZOL —————— iron-rich lower "groundwater laterite"
SOIL horizons (Buchanan's laterite)
"hardpan"
(Pendleton's laterite)

LATERITE —————— iron-rich upper
SOIL horizons — a residual soil

Pendleton's laterite was suggested (Thorpe and Baldwin, 1940) to be another form which the lower horizons of the podzolic enrichment may take, more correctly termed "hardpan" than laterite. Mohr and Van Baren (1954, pp. 356—7) described other occurrences of podzolic hardpan which are mistakenly identified as laterite.

To pedologists, therefore, laterite became something quite different from the material originally described. Nor is this without some justification, for the material was never fully defined either in terms of physical properties, chemical content, or the process by which it formed.

Hamming (1968) has reviewed modern definitions of laterite in dictionaries, encyclopedias and text books. Thus listed, their deplorable lack of unanimity is striking. He points out that laterite is still often used to mean a *type of soil*. It is certainly an intrazonal occurrence and should be distinguished from the zonal soils. The distinction proposed by Baldwin, Kellogg and Thorpe (1938) between the intrazonal Great Soil Group of the "Ground Water Laterite" and the zonal "Yellowish Brown Lateritic", "Reddish Brown Lateritic" and "Laterite" Great Soil Groups is confused by the use of the term "laterite" throughout, and there is clearly a case for a terminology for the *soils* which avoids this expression.* Hamming (1968) supports the terms used by the Seventh Approximation Soil Classification (1960). Certainly laterites must be distinguished from the zonal soils in which they sometimes occur if geomorphological conditions permit. If laterites are to be distinguished from soils, that is, if laterite horizons do not pertain to a particular soil group, what are laterites? We are forced to retain a rather loose definition.

Laterite is certainly an intrazonal occurrence, its development being dependent on special geomorphic conditions (and therefore ground and soil water-table conditions). It is also clear that laterites occur in different forms, Buchanan's being only one. Hamming (1968) has pointed out that if the term laterite is reserved for Buchanan's type with its special properties then other similar phenomena are left nameless. The present writer feels that the term "laterite" must be extended to include these other related intrazonal phenomena which are not necessarily suitable for brick-making. If it is restricted to such material then laterite is of very limited occurrence and there is none in Uganda! Sivarajasingham *et al.* (1962) have suggested a very catholic definition, which appears to be practical and serviceable. This is given on p. vi.

However, if the term is broadened to include such materials, which occur in different soil types and in different loci within their profiles, then again there is a need for classification. A useful elementary distinction can be made between groundwater and pedogenetic laterites (pp. 53—55, 73—74) which have different physical properties and which form in different loci within the profiles of the zonal soils. In the case of the former, the precipitates develop within the zone of fluctuation of the groundwater table; in the latter they form within the soil where alternate conditions of wetting and drying occur. This twofold classification has been found to be of practical field application and geomorphologically significant in Uganda. There a failure to differentiate between these laterites had led to some confusion. On the surface of the mesas assigned to the Buganda Surface, laterite is found enclosing artefacts (e.g. Wayland, 1932; Patz, 1965). This might suggest that the Buganda Surface laterite is a very recent development, and may even be currently forming. However, the artefacts are enclosed in a *pedogenetic* laterite, a recent "plaster" on the surface of the ancient groundwater laterite of the mesas which is most certainly "dead" (see Plate 4). A further

*Maignien (1966) provides a very useful summary of modern soil classifications.

problem encountered in Uganda was partially solved by the differentiation of these two classes of laterites. Pedogenetic laterite may occur in local patches, often underlain by fresh rock, which causes temporary locally impeded drainage (see Plate 3). It is not necessarily of cyclic significance (McFarlane, 1969). Laterite occurs at a great variety, if not all, altitudes, suggesting a multiplicity of planation surfaces, or partial surfaces. However when the pedogenetic laterites are identified and omitted, and only the ground-water laterites considered, the pattern of surfaces becomes clearer.

If "laterite" is broadened beyond Buchanan's concept to include all related intrazonal occurrences, a classification of these phenomena is certainly urgently required. A descriptive, possibly genetic or evolutionary classification would be the most useful for geomorphologists and might prove satisfactory for those of other disciplines. Chevalier (1949) differentiated between "young", "adult", "senile" and "fossil" laterites. Pullan (1967) also attempted a descriptive classification, again partly evolutionary. The type of classification proposed by Netterberg (1966) for calcretes in South Africa would appear to hold promise. He placed calcified soil, nodular calcrete, honeycomb calcrete, hardpan calcrete and boulder calcrete (reworked) in an evolutionary sequence, based on their physical characteristics. A similar, part descriptive, part genetic classification for the laterites of Uganda has been suggested (McFarlane, in press). The different physical types reflect different clay mineralogies, which in turn reflect differences in the environment of formation (particularly the water table conditions). The Ugandan classification, which has been found relevant to laterites in Kenya, Tanzania, Zaire and Central African Republic is summarized below. These laterites are illustrated in Plates 1, 2, 3, 8 and 9.

In conclusion, much of the dispute over laterite genesis appears to be unnecessary and due to a lack of appreciation of the essential differences between the various materials included in the term "laterite". Attempts have been made to overcome the problem by defining this term more precisely. They have failed to provide a more scientific basis for the use of this term and have perhaps complicated the situation further. There appear to be good practical reasons for the retention of the term with the broad definition suggested by Sivarajasingham *et al.* (1962) and for creating a more precise scientific classification to identify the various major types of laterite, related to the mode of genesis.

(a)   GROUNDWATER LATERITE   EROSION PRODUCTS

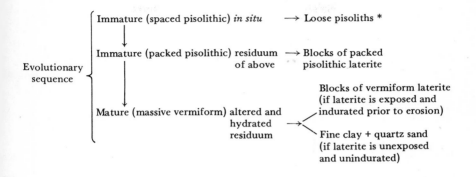

Immature (spaced pisolithic) *in situ*   → Loose pisoliths *

Evolutionary sequence

Immature (packed pisolithic) residuum   → Blocks of packed
                              of above   pisolithic laterite

Mature (massive vermiform) altered and
                    hydrated
                    residuum

Blocks of vermiform laterite
(if laterite is exposed and
indurated prior to erosion)

Fine clay + quartz sand
(if laterite is unexposed
and unindurated)

(b)   PEDOGENETIC LATERITE

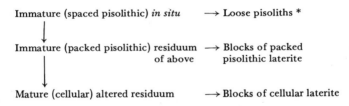

Immature (spaced pisolithic) *in situ*   → Loose pisoliths *

Immature (packed pisolithic) residuum   → Blocks of packed
                            of above   pisolithic laterite

Mature (cellular) altered residuum   → Blocks of cellular laterite

* Pedogenetic and groundwater pisoliths can be distinguished in hand specimen and clay mineralogy (Chs 7 and 8).

3

The Environment of Laterite —
Laterite and Geology

I. THE SUITABILITY OF ROCKS FOR LATERITE
FORMATION

Laterite development has often been related to what are described as "suitable rocks" (Woolnough, 1918; Hanlon, 1945) but different authors have had conflicting opinions as to what constitutes suitability. Woolnough, for example, maintained that "there is very little tendency to laterite formation on anything but the basic rock types" (Woolnough, 1918, p. 389). Simpson (1912) considered suitable rocks to be largely composed of metallic silicates, especially granite and greenstones of common types, as well as amphibolites, epidiorites, chlorite-schists and other similar rocks. Holmes (1914) noted that gneisses and basalts are particularly suitable, and Maud (1965) noted a preferential development on sandstones. Pallister (1954) observed a preference for metasediments and Wayland (1935) for igneous rocks.

Certain rocks were considered unsuitable for its development. Holmes (1914) noted its absence on granite and sedimentaries. Maud (1965) also noted its absence and Pallister (1954) its poor development over granite. Wayland (1935) recorded poorer development over metasediments and entire absence over quartzites.

Only the suitability of basic volcanics appears to be uncontended, while the absolute unsuitability of quartzites (Wayland, 1935) has been denied (Bissett, 1937; de Swardt, 1964; Hepworth, 1951; McFarlane, 1969).

In fact it appears that laterite and also bauxite development is not confined to certain rocks, but occurs over a great variety if not all rock types (McGee, 1880; du Preez, 1949; Mennell, 1909; Mulcahy, 1961; Scrivenor, 1933; de Weisse, 1954; Walther, 1915; Gozan and Vera, 1962; Blondel, 1954; Campbell, 1917; Alley, 1970; Grubb, 1963; Mabbutt, 1961). In order to explain this, it has been suggested that laterites which owe their accumulation to long-to-medium distance lateral or vertical supply have been included in these generalizations (Goudie, 1973). Although this may explain some cases of an apparent discrepancy between crust and parent material it does not provide a general explanation of the many observed anomalies.

These anomalies are informative, for *if in one area laterite development is favoured by a particular lithology while in another its development is apparently inhibited by the same lithology, this suggests that other environmental factors are operative if not dominant in its development.* This is supported by the further incongruities in the reports of the variation in the nature of laterite with lithology, particularly on an interregional scale.

II.   INTERREGIONAL VARIATIONS IN THE NATURE OF LATERITE WITH LITHOLOGY

Harrison (Harrison and Reid, 1910) was perhaps the first to suggest an apparent relationship of aluminous laterites to basic rocks, and it was soon widely believed (Scrivenor, 1937) that kaolinitic iron clays and bauxitic laterites were end-products of laterization peculiar to acidic and basic rocks respectively.

Nevertheless, although supported by many reports of the association of acidic rocks with iron clays (Buchanan, 1807; Scrivenor, 1933, 1937; du Preez, 1954; Babington, 1821; etc.) and basic rocks with bauxite (Craig and Loughnan, 1964; Evans, 1910a; Hanlon, 1945; etc.), this generalization was soon shown to be erronious. Humbert (1948) reported formation of kaolinite from both granite and basic rocks. Scrivenor (1937) drew attention to Anderson's report of bauxite development over a range of rocks including granite. Grubb's account (Grubb, 1963) of the bauxites of Jahore, Bintan and Sarawak confirms this. Fisher (1958) emphasized that bauxite was not to be associated with only the fairly aluminous rocks as was generally supposed (Evans, 1910a). He noted its occurrence over granites and granitized sediments, commenting that the variation with lithology was in iron content rather than in the degree of desilicification, or free alumina content.

These and other reports of bauxitic laterite overlying acid rocks are paralleled by reports of iron-rich clays over basic rocks (Voysey, 1833; Mennell, 1909; Dey, 1942). There is no doubt that such a heavy generalization as to ascribe the development of bauxitic laterites to basic rocks and iron-rich clays to acid rocks is incorrect. Goldman and Tracy (1946) pointed out that since bauxite can develop from kaolinitic parent material, any rock from which a kaolinitic clay can be formed may be regarded as a potential source rock for bauxite. No doubt where the other factors are favourable for development of bauxitic laterite as opposed to laterite with kaolin, then the occurrence of aluminous parent material will greatly facilitate the bauxite accumulation. Similarly for iron accumulation. However, the fact that the same rocks weather to different materials in different areas, and that quite different rocks weather similarly, certainly implies that the other environmental factors are of greater significance to the nature of the end-product than is the lithology. The apparent lack of a direct relationship between laterite and lithology was fuel to the argument that laterite is not a residuum, but a precipitate from enriched groundwaters. However, the relationship is often to be seen by less obvious accumulations than iron or alumina, for example the trace elements or heavy minerals (Ch. 9), so that the apparent incongruities between the iron or alumina contents of a parent material and the overlying laterite are no grounds for dismissing the thesis that laterite is essentially a residuum.

III.   INTRAREGIONAL VARIATION IN THE NATURE OF LATERITE AND LITHOLOGY

On an interregional scale, the variations of laterite with lithology suggest that environmental factors other than geology have had a more significant influence on the nature of the end-product. On an intraregional scale the lithological control is more evident, for within a limited area the other environmental influences such as climate and vegetation might be expected to be less variable. Thus, Holmes (1914) noted that the ferruginous laterites common on the gneisses in Mozambique do not develop over granites. These are characterized by bauxitic laterite or kaolinite. Evans (1910a) noted that the iron content of laterites could often be related to the parent material. Similar relationships have been noted by Fisher (1958), the Warths (1903) and Alexander and Cady (1962).

Often the intraregional variation of laterite with lithology is more apparent as differences in the appearance of the material. For example, Gozan and Vera (1962) noted form and colour differences in the laterites over different basalts and commented that the content was remarkably similar despite this. Du Preez (1954) noted that concretionary laterite occurs on basement complex and meta-sediments, but that oolithic and pisolithic laterites are confined to the sediments. Goudie (1973) reported that in Nigeria, Uganda and Zambia laterites on crystalline basement rocks are nearly always concretionary and vesicular whereas those on limestone and dolomites are oolitic or pisolitic. (The validity of this generalization is doubtful since there appear to be no limestones and only a very small area of dolomites in Uganda. Furthermore, pisolitic forms are extremely common on basement rocks.) Faniran (1971) noted that over shales the laterite is softer and more massive than over sandstone where it is more pisolithic. Sometimes it is reputedly possible to map the underlying geology by the variations in the appearance of the overlying laterites, so consistently is the variation of form related to lithology (R. Macdonald, personal communication).

How direct is this lithological control? The variations in the structures of the different rocks, and the way these respond to weathering, introduce microenvironmental controls in the form of differences of porosity and freedom of drainage. Thus, Pendleton and Sharasuvana (1946) noted that certain laterite structures (the concretionary structures) appear to form more readily in coarse grained and freely drained rocks than in finer grained rocks. Grubb (1963) further emphasized the role of permeability in determining the nature of the lateritic crust in Jahore. Therefore such observations as those of Gozan and Vera (1962) that the laterite varies over different basalts, and Faniran's (1971) variations over shales and sandstone, might be as readily explicable in terms of the different microenvironments these produce as by other more direct geological controls such as the chemical content.

Similarly on a slightly larger scale, variations in lithology cause variations in topography, and so an apparently direct lithological influence on laterite development might equally reflect the different topographies and their environmental variations. For example, Blondel (1954) recorded more iron-rich laterites on the higher parts of a terrain and poorer in the lower. Since it is usual for the higher parts of a terrain to be occupied by more resistant rocks, such a pattern of variation might appear to relate directly to lithology. However, it is equally possible that the lithologically-induced topography is itself the major factor.

The directness of the lithological control on an intraregional scale is certainly debatable since the geology to some extent controls the local environment; for example, the internal drainage of the saprolite, or the topography or even the vegetation of an area.

In short it appears from this review that the reported relationship of laterite to lithology is neither so direct nor so consistent as has sometimes been suggested. Where an obvious relationship exists, for example between the iron or alumina content of a laterite and that of the parent rocks, it appears to be largely a case of the lithology being such as to support the influence exerted by the environment. The obvious examples are the enormous deposits of bauxite where the environment was conducive to alumina concentration, in an area of alumina-rich parent material. More often the relationship between laterite and lithology is less clear and in many cases obscure. Nevertheless, a relationship can usually be traced, even if only in the trace elements or heavy minerals. In general a stronger influence is exerted by the environment than the lithology, particularly on the iron and alumina content. This is clear on an interregional scale and even on an intraregional scale variations may in fact reflect differences in the local environment rather than strictly geological influences.

IV. SOME EXAMPLES OF THE INDIRECT INFLUENCE OF LITHOLOGY ON LATERITE IN UGANDA

It has been suggested (Pallister, 1954) that the laterite of the Buganda Surface was less well-developed over granites than meta-sediments. The suggestion was largely based on the fact that fewer mesas survive in granite areas. Yet in Buganda the largest areas of surviving mesas occur over quartzites which contain less iron than the granite, and therefore do not appear particularly favourable to laterite development.

The survival pattern reflects not the original degree of develop-ment of the laterite but merely the resistance to weathering of the underlying rocks and the consequent ease with which lower surfaces developed at the expense of the higher. Detailed examination of the high-level laterites on granites has failed to show that they differ from those on quartzites and metasediments. Certainly the develop-ment of groundwater laterites in Buganda is independent of lithology. Other environmental factors dominate, and an interesting example of the relative roles of geology and other environmental influences can be provided from the same region by comparing the laterites of Buganda and Busia (see Location Map, p. 111).

At Busia extensive areas of low-level pedogenetic laterite have developed over amphibolites. Pedogenetic laterite, in appearance, clay mineralogy and chemical content, is quite different from the groundwater laterites of the Buganda mesas (McFarlane, 1971) and one might conclude that the differences in the laterite are caused by the lithological differences. An examination of the pedogenetic laterite profile at Busia (Fig. 1a) shows that there is a progressive loss of Ca, K and Na upwards in the profile (McFarlane, 1969). The presence of these inhibits the formation of 1 : 1 clays and the iron contained in the 2 : 1 clay minerals which predominate in the profile is not available for removal or precipitation. Only in the uppermost horizons where there is extensive kaolin formation is iron released, precipitated and accumulated as laterite. The groundwater table oscillates far below, where 1 : 1 clays predominate and no laterite can form. Thus *the nature of the lithology here dictates that it is only possible for pedogenetic laterite to develop.* In Buganda the more acid rocks allow kaolin to develop more rapidly and the groundwater table oscillates in material in which the iron is available

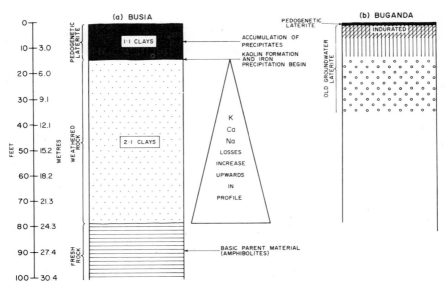

Fig. 1. *Identical pedogenetic laterites on different parent materials.*
The laterite at Busia is developed over basic parent material and in Buganda over an old, indurated groundwater laterite. The influence of the parent material is indirect; in both cases the laterite is pedogenetic because the parent material has restricted the zone of formation to the upper soil horizons. The cause at Busia is the persistence of K, Ca and Na into the upper parts of the profile and at Buganda it is the presence of the indurated groundwater laterite at shallow depth.

for solution and precipitation. Groundwater laterite therefore develops freely in this area. Nevertheless, pedogenetic laterite also occurs in Buganda as small patches overlying old indurated ground-water laterite (Fig. 1b) or rare surface exposures of fresh rock. *In both these cases laterite formation is forced into the upper horizons of the soil and a pedogenetic laterite develops.* These pedogenetic laterites are *identical* in form, clay mineralogy and chemical content to those on the amphibolites of Busia. That is, the pedogenetic laterite owes its nature to environmental factors other than geology for it occurs over a variety of parent materials. Had the other environmental factors which allow laterite to form in the upper horizons been unfavourable (e.g. climate — see Ch. 5), then there would be no laterite at all at Busia and only groundwater laterite in Buganda. Geology can thus be seen to play its part, but it is a role nevertheless subsidiary to the other environmental factors which dictate the ultimate nature or occurrence of a laterite.

V.   GEOMORPHOLOGICAL IMPLICATIONS

Although the nature of the laterite, especially its iron and alumina contents and the degree of desilicification, are not essentially attributable to geological controls it is nearly always possible to show some direct relationship between *in situ* laterite and lithology (e.g. accumulations of trace elements or heavy minerals — see Ch. 9) and this is significant to the geomorphologist. The concept of laterite as independent of lithology had supported those who followed Campbell (1917) in believing it to be a precipitate forming only after a planation surface has developed. From this stemmed the belief that a single patch of laterite surviving on an old landscape indicates that it was planation surface *par excellence.* The apparent independence of laterite from lithology was seen to accord with the accepted independence of planation surfaces from lithology. *If, however, laterite can be related to lithology* (and a secondary if not primary role is clearly to be seen), *then laterite must develop with the landsurface and not merely after it.* This was the belief of the early workers who regarded it as a residuum. If this is the case, then a patch of surviving laterite might not indicate a planation surface but merely a developing surface, or even a low relief facet which may be lithologically controlled and of no cyclic significance. Here the geomorphologist asks how do we know which it was, and do the laterites on these different topographies differ so that we can ascertain what type of surface the patch of laterite represents?

4

The Environment of Laterite — Laterite and Topography

Although laterite has long been associated with low relief (Buchanan, 1807, p. 460; Blanford, 1859; Oldham, 1893), and occurs frequently on level or near level surfaces (Maclaren, 1906; Campbell, 1917; Wayland, 1931; Pendleton and Sharasuvana, 1946; de Vletter, 1955; Mabbutt, 1961; Sombroek, 1971), very few of the early workers believed that its formation was *restricted* to level surfaces. It was believed to be residual and the necessary conditions were visualized as merely a landsurface of sufficiently low relief to allow appreciable ingress of surface waters. Steep slopes cause rapid run-off and indiscriminate removal of weathering products, while shallow slopes allow an ingress which causes the differential removal of the more soluble constituents and the accumulation of the less mobile weathering products. This idea was established before the concept of cyclic development of landsurfaces, and at one time it was believed that the laterite on plateaus was forming contemporaneously with the low-level laterites, simply because these two low-relief situations existed (for whatever reason), albeit altitudinally separate. Although Newbold (1846a) had already expressed the opinion, it was Woolnough (1918) who established the belief that the upland laterites are "dead" and being eroded.

The concept of laterite as a residuum fitted well with the "normal" Davisian cycle of landsurface development (Davis, 1909, 1920) when this became established. During the reduction of a landsurface slopes are reduced until they reach a critical value so that ingress is substantially greater than run-off, and laterite genesis begins. Laterite development was thus believed to be associated with a stage of landsurface development in which the relief is still being reduced. The laterite develops *with* the landsurface and the end-product is a planation surface blanketed by a thick sheet of residual material.

It is important to distinguish between this early concept of laterite developing with the landsurface and the later concept of laterite as a precipitate which distinctly implies that laterite formation postdates the formation of the planation surface. In this theory it was only *after* the peculiar conditions of a stable and fluctuating water table had been created by the formation of the planation surface that laterite genesis could begin. Thus, Woolnough (1918, p. 390) noted that "laterization can occur only in areas where drainage is almost at a standstill. This usually involves the existence of a peneplain almost at sea-level". Maclaren (1906) had also concluded that "only on a level surface can laterite form". Wayland (1931) similarly subscribed to the belief that "it is of course *consequent* upon peneplanation" (present writer's italics). The concept of laterite as a precipitate has therefore the result that it restricts it to planation surfaces in a way that the earlier concepts of laterite as a residuum did not. A thick sheet came thus to be regarded as indicative of a planation surface which *survived as such* for a considerable length of time. The very existence of a laterite thus assumed paramount importance for studies in denudation chronology.

However, although laterite is commonly associated with low relief it is by no means confined to it (Goudie, 1973, p. 42). It occurs on mature (Fisher, 1958) and submature (Nye, 1954) landsurfaces and has recently been shown to occur on landsurfaces with considerable relief (Stephens, 1961; Johnson, 1954, 1959; Pallister, 1951). Up to 500 ft (152 m) of relief has been recorded on continuous sheets of laterite (Johnson and Williams, 1961; Trendall, 1962; Hepworth, 1951, 1952a,b), and de Swardt (1964) described the Buganda Surface as an uneven surface of high relief in the order of 600—900 ft (183—274 m).

Moreover, laterite is apparently not restricted to shallow slopes: 7—10° being frequently recorded (Pallister, 1951, 1954; Trendall, 1962; Hepworth, 1952b; Mulchy, 1960; Alley, 1970; Playford, 1954). De Swardt (1964) described laterite on slopes of up to 20° and Hepworth (1952a) recorded up to 22°. So high is the relief of the Buganda Surface that Trendall (1962) attempted to explain it by an entirely new theory of laterite genesis associated with the formation of an "apparent peneplain" (Ch. 10), an acyclic formation.

Clearly such relief is incompatible with the concept of laterite as a precipitate and it is difficult even to reconcile it with the concept of laterite as a residuum. It has been suggested that this high relief is not the original form of the surface but is due to postincision modifications. Three types of modification have been suggested.

1. *Local cambering of the margins of mesas* (Pallister, 1953)

Local cambering is the result of basal sapping and peripheral sagging of the carapace (Fig. 2a). Certainly in Uganda this type of deformation can be clearly seen. It cannot, however, explain the total relief of up to 500 ft (152 m) or more on long sloping ridges.

2. *Accelerated erosion of the margins of mesas to form a "waxing slope" above the laterite scarp* (Pallister, 1953)

A waxing slope above the laterite scarp may develop by the erosion of overlying soil material so that the laterite originally found well down in the soil profile is exposed at the periphery of the mesa while the centre still retains the original thickness of overlying soil material (Fig. 2b). Although erosion of overlying material may be accelerated towards the margins of a mesa, the records of laterite outcropping *on the surface* at close intervals if not continuously from the scarp edge to the summit (Hepworth, 1951, 1952a; McFarlane, 1969) are numerous and this explanation is quite inadequate to account for the observed relief. Certainly the assumption that laterite normally occurs overlain by hundreds of feet of soil is unsupported by evidence. The development of the waxing slope could also be due to erosion of the margins of a thick sheet or body of laterite and the truncation of the profile (Fig. 2c). Hepworth (1951, 1952a) discussed this possibility and dismissed it on the grounds that there is no evidence for bodies of sufficient thickness to explain hundreds of feet of surface relief in terms of profile truncation.

3. *Bicyclic mesas* (McFarlane, 1969)

An examination of the originality of the relief of the high mesas in parts of Buganda has shown that a single mesa may in fact be bicyclic. Two chronologically separate laterites often occur linked by a detrital laterite, the product of erosion of the higher element (Fig. 2d). This detrital laterite overlaps the lower and younger laterite forming a continuous sheet (see Plate 17). This can account for surface relief of 500 ft (152 m) or more on a given mesa.

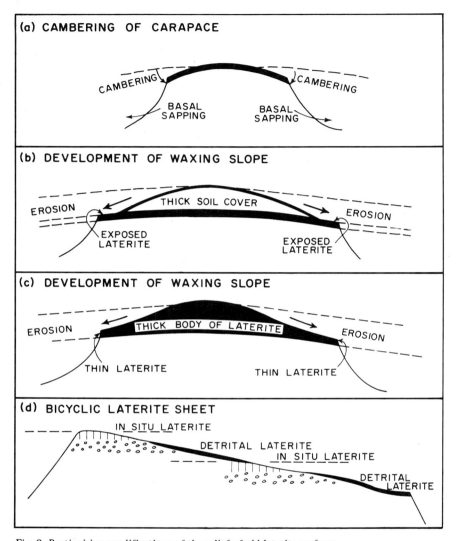

Fig. 2. *Postincision modifications of the relief of old laterite surfaces.*

The high relief of the high-level laterite mesas must certainly be regarded with circumspection since they are extremely ancient features and much modified. It is certainly quite incorrect simply to draw a line round the scarp edge of a mesa and assume that within this the existing relief must represent a single surface in its original form.

The low-level laterites, being largely unincized and unmodified, provide safer data on the original relationship of laterite to topography. Certainly most of the reports of incongruously steep slopes and high relief concern the incized laterite sheets. Nevertheless it is quite clear that the low-level laterites in many cases are by no means flat and where there is undeniable relief this argues against the concept of laterite as a precipitate formed consequent upon the formation of a planation surface.

The variations of laterite with topography have received relatively little attention compared with the variations with lithology. The earliest recognized and most familiar variation is the often described differences between high- and low-level sheets. As early as 1893 Oldham discussed the already accepted concept of high-level *in situ* laterite and low-level detrital laterite, and this distinction has been generally adhered to (Pallister, 1954; Wayland, 1935; Johnson, 1959; etc.). It has the implication that these sheets form in entirely different ways, the lower by re-solution of detritus from the *in situ* laterite. Although there is no doubt that low-level laterites are enriched from the high-level laterites (Maignien, 1958; d'Hoore, 1954; McFarlane, 1969) a review of the relevant literature clearly shows that the distinction is more firmly entrenched as a concept than is warranted by the evidence. In concluding his review Oldham (1893) considered that "no hard and fast distinction can be drawn between high- and low-level laterites". This opinion appears to have gone largely unnoticed, as has the observation that Buchanan's laterite is in part detrital (Oldham, 1893; Scrivenor, 1933). King (1882) even suggested that the detrital variety should be called laterite and the *in situ* variety lateritized rock.

Studies of low-level laterites such as that of du Bois and Jeffery (1955) have clearly indicated that there is an *in situ* element of low-level laterites worthy of study in its own right, albeit recognizing the existence of foreign surface additions. Recent investigations in Uganda have confirmed this, using differential thermal analysis. Pisoliths enclose and effectively fossilize the material in which they form (Ch. 8). Analyses of the sheets of low-level pisoliths have shown that they enclose quantities of 2 : 1 clay minerals (McFarlane, 1969) which do not occur in the pisoliths of the high-level sheets. These latter formed in extremely weathered saprolite characterized by 1 : 1 clay minerals and so the pisoliths containing 2 : 1 clay minerals cannot have been inherited from the older sheet. The low-level sheets are therefore at least in part "new" or "first order" laterites (see Plates 10 and 14).

The distinction between high-level *in situ* laterites and low-level detrital laterites is difficult to justify (compare Plates 9 and 14) and even the distinction between *in situ* and detrital laterites does not bear close inspection. For example, Nye (1954) described a laterite catena in West Africa, in which laterite nodules collect in the base of a down-wasting soil and are redissolved to form a pan in lower topographic positions. Similar situations occur in Uganda (McFarlane, 1969). The re-solution of a residual sheet of pisoliths which have down-wasted through a few hundred feet to their present position and the re-solution of pisoliths which are detrital in that they have fallen a few hundred feet down a subcarapace slope are similar pan-forming processes. In the latter case the laterite is clearly detrital, but the former is debatable. The laterite is not truly detrital in the sense that it has inherited material from an older crust. It is a "new" or "first order" laterite, but to call it *in situ* would be misleading. It is questionable whether, if a laterite is identified as residual, it can be regarded as *in situ* at all, since vertical and lateral migration of particles is an inevitable part of its formation (see Plate 10).

Thus, a review of the literature and evidence in Uganda both fail to demonstrate the popular distinction between high-level *in situ* laterite and low-level detrital laterite: furthermore the basic distinction between *in situ* and detrital laterite is far from clear.

Less familiar than the reputed high- and low-level variation of laterite is the variation within a single sheet. From the literature available it would appear that laterites characterized by mottles, pisoliths and nodules are associated with a higher degree of relief than are reticulated or vermiform laterites. Thus, Nye's laterite (Nye, 1954), occurring on a rolling topography, is characterized by pisoliths and concretions. Radwansky and Ollier (1959) recorded formation of concretions in the C and B horizons of a soil where slopes are in the order of 9 and 10°. Wood and Beckett (1961) recorded mottle formation at depth on 15 to 20° slopes. Mulcahy (1960) described mottles and concretions formed in valley sides with slopes of up to 8°. Moss (1965) noted that the mottled clays of south-west Nigeria are not confined to a narrow altimetric zone but may occur on long convex slopes extending through a vertical range of 200 ft (61 m) and added that while mottled ferrallitic clays are rarely associated with slopes of more than 4°, the non-mottled ferrallitic clays are found associated with slopes of 1° or less.

Reticulated clays are usually associated with topographies that are almost flat (Mohr and Van Baren, 1954; Thorpe and Baldwin, 1940).

Where both concretions and reticulated clays occur together in one monogenetic unincized profile the former have been found to overlie the latter (e.g. Ollier, 1959). The formation of mottles, concretions and pisoliths may thus be associated with a greater degree of aeration and freedom of drainage than is that of the reticulated clays. Alexander's observation (Alexander and Cady, 1962) that pisoliths form early in a given laterite supports this. Humbert in 1948 had already suggested that the concretion stage is a definite step in the genesis of laterite, and there seems little doubt that such structures characterize *immature* laterites which often form on surfaces of low but nevertheless significant relief. The pans, reticulated clays and massive laterites are mature and more properly planosol developments.

This relationship is clearly demonstrable in Uganda. Low-level laterites occur on a rolling topography. On the higher, more freely drained areas, which are still actively down-wasting, immature laterites characterized by pisoliths and other discrete concretions occur (see Plates 10 and 14). The lower parts of the catenas are characterized by a more mature massive vermiform laterite, where down-wasting has virtually ceased and the drainage is poorer. It has thus been possible to classify the laterites of Uganda in a part descriptive, part genetic classification; for the different laterite structures, indicating the degree of maturity of the laterite, can be related to the degree of maturity of the surface on which they occur (McFarlane, 1971 and in prep.). Such a classification is clearly applicable in geomorphological studies. It has been applied in Buganda to analyses of the so-called Buganda Surface where, when the different types of mesa laterites were mapped and plotted altitudinally, the mesas were found to comprise two planation surfaces, each marked by a massive vermiform laterite (McFarlane, 1971, also Plate 8). Pisolithic laterites (Plates 1 and 2) occurred altitudinally between these two planation surface laterites (see Plate 17). That the vermiform laterite occurs on the highest interfluves shows that the older of the two surfaces, which we may call the Buganda Surface proper, was clearly a planation surface *par excellence*. This study shows the futility of mapping the laterite as such without considering the variations of laterite with topography.

Summary

The relief of many high-level laterite surfaces may be due in part to postincision modifications of the incized carapaces. However, this

37

does not provide a general explanation, and the existence of laterite surfaces of appreciable relief argues strongly against concepts of laterite as a precipitate which restrict its development to planation surfaces. Such evidence favours the concept of laterite as a residuum developing with the surface. Concerning the variations of the nature of laterite with topography, the reputedly essential difference between high-level *in situ* and low-level detrital laterites is in practice far from clear, as is the general distinction between *in situ* and detrital laterites themselves. Less well-known are the variations within a single sheet. Since laterite structures vary with topography this may provide the geomorphologist with a useful tool; for when the relationship of laterite type and topography has been established it becomes possible to use the nature of the laterite to assess the status of the old, modified, laterite mesas, that is, to determine whether or not they are significant as "markers" in the denudation chronology. Certainly every patch of laterite does not represent a planation surface.

5

The Environment of Laterite —
Laterite and Climate

I. TEMPERATURE REQUIREMENTS

The early accounts of laterite genesis implied that it could not form in extratropical environments (Newbold, 1846b), and certainly the belief that hot or warm environments favour its formation is widely popular (Fisher, 1958; d'Hoore, 1954; Sandford, 1935; de Vletter, 1955; Sivarajasingham *et al.*, 1962; du Bois and Jeffery, 1955; Humbert, 1948; McNeil, 1964; du Preez, 1954; Nye, 1954; Doyne and Watson, 1933; Maignien, 1958). There is less unanimity of opinion about the incidence of this heat. Pendleton (1941), for example, emphasized that continuity of high temperatures is favourable, while McNeil (1964) believed that high temperatures are only necessary "at least during part of the year", and Van der Merwe and Heystek (1952) described current laterite formation where there are "high" summer temperatures and "mild" winter conditions. A minority believed that it may also form in temperate environments (Harrison and Reid, 1910; Scrivenor, 1910b; Woolnough, 1927; Simpson, 1912; Prescott and Pendleton, 1952).

When improved techniques enabled the individual components of laterite to be identified, the conditions favourable for the development of these individual constituents could then be investigated. In particular the conditions favourable to the formation of alumina and kaolin and for the development of low silica sesquioxide ratios promised a contribution to the understanding of the temperature requirements for laterite formation. However, studies of the effects of temperature on the development of these components have yielded rather confusing results. Krauskopf (1956) showed that silica solubility is linearly related to temperature, and this suggested that desilicification, widely believed to be synonymous with lateritization, is aided by high temperatures. Jenny (1929) had indicated this by showing that the SiO_2/Al_2O_3 ratio falls with increasing temperature. There was little agreement beyond this. Robinson and Holmes (1924) using other methods, found no agreement between temperature increase and degree of lateritization as expressed by this ratio. Crowther (1930) disputed the validity of Jenny's calculation and attempted to show that the relation was not, as he suggested, negative, but positive, that is, the ratio increases with temperature, the clays becoming more siliceous. Russell (1962) in turn suggested that Crowther's results were based on inadequate methods, and concluded that the effect of temperature on the formation of particular clays is uncertain.

It is, however, generally believed (Russell, 1962) that the higher the temperature of percolating water the more effective it is in decomposing the rocks and lowering the silica content.

II. RAINFALL REQUIREMENTS

It is generally accepted that rainfall conditions are more critical to laterite formation than temperature conditions but prior to the development of the concept of laterite as a precipitate there was little discussion of the rainfall conditions required. It was assumed that water must be in abundance. About 1910, opinions began to diverge. Maclaren (1906) stated that a well-marked alternation of wetting and dessication is essential. Mennell (1909) agreed with this. There was an immediate repudiation of this from Scrivenor (1909) who stated that seasonal rainfall is not necessary and gave rainfall figures for Malacca where laterite was believed to be currently forming. Thenceforth the field became fairly split between those who considered marked seasons to be essential and those who believed them to be unnecessary.

As d'Hoore (1954) pointed out, water, in some quantity, is essential. As soon as this medium of transport is absent there is no accumulation. It is also generally agreed that *alternating conditions* are needed for sesquioxide precipitation. This readily came to be regarded as synonymous with alternating *wet* and *dry* conditions. Wet and dry *seasons* came thus to be considered favourable if not essential to laterite genesis (Martin and Doyne, 1927; du Preez, 1954; Fisher, 1958; Humbert, 1948; Simpson, 1912; Sabot, 1954). The great depth of some laterite profiles caused Walther (1916) to suggest that sufficiently extreme conditions no longer exist on this planet. Most, however, believed laterite to be explicable in terms of currently occurring seasonal climates. Some favoured seasons of equal length (Alexander and Cady, 1962; Sivarajasingham *et al.*, 1962), while others considered such regularity unnecessary (Humbert, 1948; Woolnough, 1927; Panton, 1956) or even unfavourable (Maignien, 1958). Nevertheless there remained a strong body of opinion which favoured entirely humid conditions (Scrivenor, 1909; Harrison and Reid, 1910; Jensen, 1914; etc.). Many writers, although accepting that alternating seasons facilitate laterite formation, considered this to be unessential (Mohr, 1944; Fisher, 1958).

There are several ways in which such a marked divergence of opinions may be reconciled. Climatic change may have occurred. The current conditions are not necessarily those prevailing at the time of formation of the laterite. Goudie (1973) has emphasized the importance of palaeoclimates in the study of duricrust-forming conditions and cited some Australian studies (Dorman and Gill, 1959; Gentilli, 1961; Gill, 1961) indicating the considerable changes which have occurred since the Tertiary, the reputedly major duricrust formation period. Thus the prevailing conditions in a laterite area do not provide satisfactory data concerning conditions favourable to laterite genesis, and it is often very difficult to ascertain whether or not a given laterite is actually currently forming.

A more promising means of reconciling opposing views lies in recognizing the undue stress placed on atmospheric climate as opposed to what Maignien (1958) has described as "climat du sol". This distinction, discussed by Mohr and Van Baren (1954), is very important. Thus, for example, although the laterite Nye (1954) described occurs where there is a three- to six-month dry season, there is a total rainfall of 50—150 in (127—381 cm) and a forest vegetation. Doyne and Watson (1933) described laterite formation where there are "marked seasons", and added that the natural vegetation is forest. Clearly to stress the atmospheric climate is deceptive, for the climate of the soil supporting a true forest vegetation cannot be markedly seasonal; and it is surely the soil climate which is more significant to laterite genesis.

Another way of reconciling these opposing views is by resolving the common confusion of pedogenetic and groundwater laterite (pp. 53—54, 73—74). It is clearly important to determine the horizon in which a laterite develops; for the "climate of the soil" (the upper horizons) need not be the same as the "climate" of the water table zone, much as the "climate of the soil" need not directly reflect the climate of the atmosphere. Thus the oscillation of a water table will provide alternating conditions in both a seasonally wet and dry regime and in a permanently wet climate where the rainfall regime is uneven. Water table oscillations will occur in even the most regular regime if only over periods of several years. Thus, alternating conditions are provided at this level in any climate and it may be true to say that groundwater laterite may form in *both* permanently wet and seasonally wet and dry regimes (Fig. 3). For pedogenetic laterite to form, alternating conditions must be provided in the upper soil horizons and thus this kind of laterite may be best formed in an unforested environment where seasons are marked (Fig. 3a). Its formation may be completely inhibited where the rainfall is evenly distributed (Fig. 3c).

These suggestions are supported by the evidence in Buganda. Here the groundwater laterites appear to have formed under a forest vegetation (McFarlane, 1969). The natural vegetation today is forest,

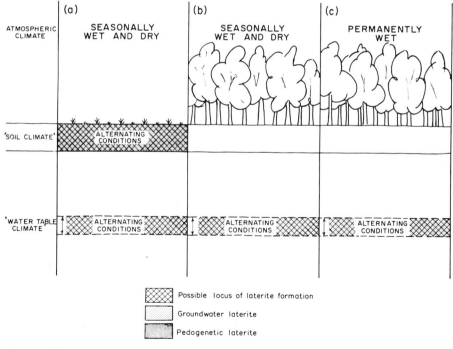

Fig. 3. *"Climate" and the locus of laterite-forming conditions.*

although the climate is classified as seasonal. The dry season is not, however, sufficiently marked to inhibit forest growth. The trees effectively buffer the soil against the variations of the atmospheric climate and the "climate of the soil" must be regarded as permanently wet. It is noticeable, however, that where man has destroyed the forest vegetation, causing it to be replaced by grass, local patches of relatively recent pedogenetic laterite have developed and are still developing. This is well seen on the Buganda Surface, for where the forest is destroyed and the soil eroded, the old groundwater laterite has become indurated and impermeable (Fig. 4 and Plate 6). The seasonal nature of the atmospheric climate is exaggerated at soil level. The soil becomes alternately saturated and baked dry. The "climate of the soil" is in effect more markedly seasonal than the atmospheric climate and the development of pedogenetic laterite is favoured (see Plate 4).

Thus, it is important in discussions of the climate of laterite formation to specify the locus of the laterite formation and to recognize that alternating conditions (necessary for sesquioxide precipitation) are not to be entirely equated with an atmospheric climate with alternating wet and dry seasons.

Although they do not provide information on the incidence of the rainfall, studies of the effect of humidity on the formation of the individual laterite constituents were more fruitful than the studies of

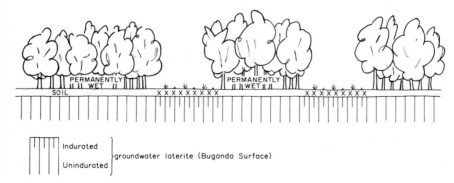

Fig. 4. *Deforestation and the creation of new laterite-forming "climatic conditions".*
Beneath forest-cover permanently moist soil conditions obtain. Deforestation exposes the underlying groundwater laterite to seasonal variations causing induration and loss of permeability. The seasonality of the atmospheric climate is also felt at soil level and pedogenetic laterite develops as a veneer over the groundwater laterite.

the effects of temperature on these constituents. Martin and Doyne (1927) first pointed out that the less the rainfall, the higher the silica sesquioxide ratio; that is, the less the degree of lateritization. More recently, Van der Merwe and Heystek (1952) showed that lateritization, as measured by the progress of illite to kaolinite and to bauxite, increased with rainfall. Crowther (1930) likewise found the SiO_2/Al_2O_3 ratio to decrease with increasing rainfall. He also claimed that Jenny's temperature data (Jenny, 1929; also see p. 40) in fact showed this relationship of laterite development to rainfall. Thus, there is fairly general agreement that the higher the rainfall, the better are laterite-forming conditions.

These conclusions are supported by consideration of the whole laterite profile, not merely the laterite itself. Typically, groundwater laterite is underlain by very deeply weathered profiles, for example Nagell (1962) cites 100 m of weathered zone; and depth of weathering is directly related to moisture available (Nye, 1954; Van der Eyk, 1965). Clearly a climate with permanently wet conditions will be more conducive to deep weathering than a climate in which moisture is only available for part of the year (see Plate 13). Thus, the postulate that seasonal moisture is most favourable to laterite formation creates the incongruous situation where part of the typical profile is best developed under permanently moist conditions and part under a seasonal regime. For example, Woolnough (1927) stated that where laterite was found to overlie a deeply weathered residual profile, this indicated a uniformly moist climate followed by an alternating one. Likewise Watson (1965) considered a wet climate necessary for the formation of a deep weathering crust and "moderate rainfall" for the formation of the laterite. To postulate such combinations of climate to explain the typical groundwater laterite profile seems unwarranted. By this argument the profile beneath the laterite appears to support the contention that a permanently moist environment is the more favourable for laterite formation.

Recently, a method attempting to assess the climatic limits of laterite formation devised by Kerner-Marilaun in 1927 has aroused renewed interest (Goudie, 1973). This method considers (a) quantity of precipitation, (b) seasonality, and (c) temperature, to be critical variables and, based on empirical evidence, it is proposed that laterite will form where L exceeds 50 in the formula:

$$L = R^{-\frac{1}{4}} \cdot (S - s) \cdot tm \cdot 100^{-1}$$

where L is the "laterite number"; R is the annual precipitation in mm; S is the wet season semiannual precipitation (mm); s is the dry season semiannual precipitation (mm); *tm* is the minimum monthly mean temperature (degrees C).

High *L* values, that is, favourable laterite formation conditions, would appear to be associated with a very high markedly seasonal rainfall and high temperatures. However, by this formula, the requisite *R* value for laterite formation could never be achieved in a situation where the rainfall, however high, was evenly distributed throughout the year. This method is reported (Goudie, 1973) to have met with some success in South America, but since it considers only atmospheric conditions and not those occurring at the locus of laterite formation, it is unlikely to be of very wide application, although it may possibly be relevant to certain types of laterite formation, for example the pedogenetic laterites.

Summary

Although warm temperatures are generally considered to favour laterite genesis, there is not complete unanimity as to its exclusive development in the tropics. The results of studies of the formation of the individual laterite constituents are confusing but it is generally accepted that high temperatures favour desilicification. The moisture requirements are considered to be more critical. Alternating conditions are required for sesquioxide precipitation and after the development of the concept of laterite as a precipitate this came to be regarded as synonymous with marked wet and dry seasons. However, there is also considerable evidence for its formation under permanently moist atmospheric conditions. Some of the anomalies observed in the field may be attributed to climatic change. More important is the lack of appreciation that the climate of the soil is more significant than the atmospheric climate. The buffering effect of a forest vegetation may reduce seasonal variations and the formation of many laterites would be thus more correctly associated with a less variable regime than the atmospheric climate suggests. Alternating conditions may be provided at groundwater table level in all climatic regimes, and for this reason it is important to distinguish between pedogenetic and groundwater laterite. The development of a pedogenetic laterite may be restricted to areas with a markedly seasonal atmospheric climate and "soil climate". Studies of the formation of the individual laterite constituents suggest that moist conditions are generally favourable, and the depth of weathering below the typical groundwater laterite also indicates a moist environment of formation.

6

The Environment of Laterite — Laterite and Vegetation

The theory that laterite develops in a grassland environment, first formulated by Harrassowitz (1930), is firmly entrenched. It dates essentially from the development of the concept of laterite as a precipitate forming at the upper limits of a seasonally fluctuating water table (pp. 3—6). Seasonal water-table fluctuation and a seasonal climate are more readily associated with a grass vegetation than with forest, and d'Hoore's observation (d'Hoore, 1954) that iron is more mobile under grass than forest, is in agreement with the stress laid by this school of thought on the mobility of iron as the cause of the accumulation (p. 4).

Laterite is undoubtedly *seen* more frequently in grasslands but it is by no means exclusive to them (e.g. Nye, 1955; Harrison and Reid, 1910; Nagell, 1962). The early accounts of laterite, those predating the concept of it as a precipitate, usually described it in a forest environment (Buchanan, 1807; Babington, 1821), and if it occurred in a grassland environment it was usually made clear that this was not a climatic but an edaphic climax. For example, Buchanan (1807, p. 560) noted that "where the earth has lodged sufficiently deep to retain some small degree of moisture" occasional clumps of trees are to be found on the otherwise bare or grass-covered laterite surfaces. Newbold (1846a, p. 990) also noted that "wherever there is sufficient depth of soil and capability of retention of moisture its (laterite) chemical composition is certainly not against arboreous vegetation". He suggested that the porosity of laterite may account for its reputation for sterility. Oldham (1893) supported this. This pattern of grass on the laterite and forest in gullies and depressions where soil is thicker is very common, not only in India (e.g. Frasché, 1941; McFarlane, 1969).

Many of the descriptions of such grassland communities show that the original climax, probably climatic, passed through a man-made climax before reaching the present edaphic climax. Thus, Buchanan (1807, p. 460—1) noted that the long grass was frequently burned "which destroys the bushes that spring up in the rainy season, and [this] keeps the country clear". McGee (1880) made similar observations. Such grasslands are temporary and man-made. Hardy and Follett-Smith (1931) recorded laterite under low forest of mixed composition, comprising secondary growths following the partial clearing of the primary forest by man. Clearly both the intensity and duration of man's interference governs the nature of the vegetation. If such interference is of short duration, regeneration of the original climatic climax communities may be possible. If however, the interference continues for any length of time, permanent edaphic climaxes or "prairies irreversible" (Schnell, 1949) result. Harrison and Reid (1910) described the process by which such edaphic climaxes develop:

"In the dense forest of the Guianas there may be said to be a perpetual wet season, as under the shade of the trees even during periods of comparative drought, the land is invariably wet . . . . The protective influence on the soils of the very heavy tropical forests which in the Guianas specially characterise the areas of lateritic residual deposits is very great . . . . When the land is cleared of forest, denudation rapidly removes the fine constituents of the earths, leaving on the surface the masses of ironstone, bauxite and quartz".

Thus, despite the circumstantial evidence of a seasonal climate, with periods of drought, the sparse vegetation is edaphically controlled.

The laterite mesas of Uganda are also predominantly grass-covered, standing like islands in the more heavily vegetated lowlands (see Plate 7). Nevertheless it is quite clear that they were formerly forested. Patches of forest remain in the centres of the largest mesas (see Plate 6). In many places the soil cover is sufficiently thick to support forest were this allowed to regenerate, but cultivation, burning and grazing hold the vegetation at the grass level. The result is clear. The soil is eroded and the laterite is exposed and becomes indurated and impermeable. Once this has happened, only a thin cover of grass manages to survive as a permanent edaphic climax.

47

Temporary climatic change towards drier conditions or more marked seasons may have an effect similar to man's activities, inducing permanent edaphic grass climaxes. Climatic change is a popular mechanism, freely invoked to explain the development of laterite in areas where the current climate does not conform with a particular author's opinion as to the climate suitable for laterite genesis (Sandford, 1935; Kellog, 1949; Litchfield and Mabbutt, 1962; Hepworth, 1951; Ab'Saber, 1959; Wilhelmy, 1952). However, there has been an increasing appreciation of the extent to which man is capable of altering his environment (Dey, 1942). Much of the deforestation of the laterite surfaces must have been for agricultural purposes and may date from the Neolithic. On Bugaia and Buvuma Islands on Lake Victoria (see Location Map, p. 111), ancient field boundaries and terraces are to be found on the mesas in an area where terracing is now completely unknown (McFarlane, 1968). However, much of the clearance may be considerably older. Stone artefacts are frequently found embedded in laterite or laterite rubble (Foote, 1880; Wayland, 1932; Patz, 1965; McFarlane, 1968) indicating that stone age man frequented many areas characterized by laterite. One is tempted to draw a comparison with the popularity of the chalk uplands of England for early human occupancy, since both laterite and chalk uplands provide dry sites on which forests do not regenerate rapidly once they have been cleared. The effectiveness of stone age agriculture in modifying the environment appears likewise to have been underestimated (Dimbleby, 1961; Geddes, 1960; Proudfoot, 1967), but much of the forest destruction may have been caused by fire, either accidentally or perhaps deliberately to aid hunting (cf. Sombroek, 1966, p. 187). The earliest use of fire which has been dated in Eastern Africa, in Rhodesia, occurred in the Final Acheulian, 57,000 years B.P. (Oakley, 1961), so that in Africa man has long possessed the means of radically altering his environment.

Thus, although climatic change may be responsible for the induction of permanent edaphic grass climaxes on laterite, it appears that man's role in this respect may have been underestimated.

In short, it may be said that the very frequent association of laterite with grassland appears to favour the belief that laterite forms as a precipitate at the upper limits of the range of a seasonally fluctuating water table; for grass is often the climatic climax in areas with seasonal climates. However, laterite also occurs in forest environments and many of the occurrences of grass vegetation on laterite are permanent edaphic grass climaxes or temporary man-induced grass climaxes. If such grasslands are edaphic rather than climatic climaxes, that is, the grass exists *because* it is underlain by laterite, then the very common coexistence of grass and laterite cannot be taken as evidence that laterite *formed* in a grassland environment.

Observations such as those made by Newbold (1844, p. 995, 1846a, p. 204), of beds of lignite in detrital laterite deposits and interbedded silicified wood in sandstones of the same age as the laterite, perhaps provide more significant information about the nature of the vegetation at the time of laterite formation than does the vegetation currently seen on the laterite.

The suggestion that laterite, characterized by kaolin and iron, may be expected to develop where vegetation is heavy appears to conflict with some observations that iron is not stable in such an environment, but is readily removed by humus-rich solutions (Hanlon, 1945). Goudie (1973; p. 113) has listed what he believes to be the reasons why tropical rainforest is often inimical to lateritization. Soil acidity is cited as being unfavourable. However, although acid conditions may occur near the surface it is widely known that high latitude profiles are more acid than their tropical counterparts (Curtis, 1970). In the tropics, acidity is not maintained with depth in the profile and neutral or alkaline conditions may be the norm (A. Mehlich, personal communication; also see Ch. 9). It is to this low acidity that the solubility of silica has been generally attributed in the tropics (cf. Krauskopf, 1967, p. 194). The consumption of humus by abundant microflora in the soil and the consequent reduction of the concentration of humic acids in solution is recognized as one means by which the pH of humid tropical soils is kept high. Certainly, although abundant litter is provided, its decay is rapid and forest floors are often only partly covered by a thin veneer of freshly fallen debris (e.g. Nagell, 1962). Moreover, even where acid conditions do occur, leaching tests by Pickering (1962) suggest that relative accumulation of lateritic materials can be expected. Goudie (1973) has indicated that large amounts of litter are unfavourable to laterite formation, but Bloomfield (1953) has shown that although aqueous leachates of forest litter readily dissolve hydrated ferric and aluminium oxides with the formation of soluble ferrous and aluminium complex compounds, sorption of the reaction products takes place concurrently with the solution process and the sorption has the effect of inhibiting further solution of the ferric oxide. Rodin

49

and Bazilevich (1967) have indicated the positive role played by vegetation in the relative accumulation of iron and aluminium in tropical soils. Goudie (1973) also states that

"forests tend to protect the underlying soil from dehydration (a powerful factor in the development of induration) and also exert physical forces likely to inhibit crust development".

This is an example of the common confusion of "induration" and "development" of laterite which has led to the vague general belief that forests are unfavourable to laterite formation. Forests certainly inhibit its induration, but that is not synonymous with its formation or development. Evidence to support the suggestion that iron stability is innately incompatible with a forest environment is lacking.

Further evidence of the influence of vegetation on laterite genesis is based upon observations of the destructive effects of the vegetative cover. There are many references to the ability of trees to soften, dissolve or break up a laterite carapace (Sivarajasingham, *et al.*, 1962, pp. 55—56). This has the inevitable implication that a forest growth is not conducive to iron stability or precipitation, that is, to laterite formation. However, it is essential to see this destructive ability in its proper context. Trees which are dissolving laterites are situated on surfaces which have been incised or eroded and which therefore bear unstable or truncated soil profiles. In unmodified profiles the horizons of sesquioxide accumulation occur at depth and not at the surface. If such a stable profile is disturbed by erosion of the topsoil or by lowering of the water table (which lowers the "base" to which the profile is "graded") the resultant instability of the profile may be the cause of re-solution of the precipitates as the profile adjusts to a new state of "grade" in the altered circumstances. With removal of the topsoil, the horizons in which the sesquioxides are stable will be brought into a relatively higher position in which they are innately unstable, and will be dissolved (Fig. 5a,b). They may be reprecipitated in the new (B) horizon if the loss of stability was only caused by slight erosion of the topsoil. If, however, due to major incision, the water table is considerably lowered, then reprecipitation may not occur and the dissolved material may be entirely lost to that particular profile (Fig. 5c). Bisset (1937) described in Uganda an enclosed hollow on a laterite mesa, which he excavated to find no laterite at the base. This would appear to be an example of such complete loss of the precipitates as a result of the loss of stability in

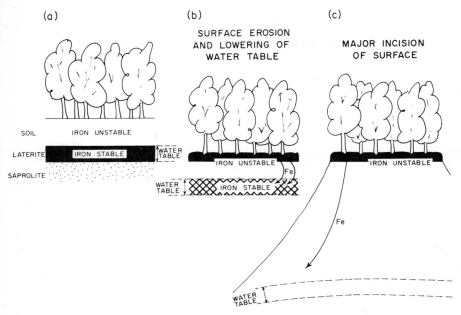

Fig. 5. *Re-solution of laterite under a forest cover.*
(a) and (b). Soil erosion brings the laterite horizon relatively higher in the profile, into a zone of inherent instability, and solution of the laterite occurs. Deposition lower in the profile may allow laterite to re-form if the disturbance is minor.
(c). If the disturbance of the profile is major, then the constituents dissolved may be lost from that profile and the laterite gradually destroyed.

the profile caused by water table lowering. Clearly therefore the re-solution of *eroded* laterite profiles carrying a forest cover is not sufficient evidence to postulate an innate incongruity between laterite genesis and forest vegetation. Indeed, if such a process of re-solution of laterite is further considered it can be seen to contribute positively to the understanding of laterite genesis, for it provides a possible mechanism for the *accumulation* of the precipitates in situation where the water table is progressively lowered and there is gradual complementary topsoil removal. In such circumstances, as would be provided by a gently down-wasting landsurface, there would be a continuous process of solution of the precipitates at the upper limit of the zone in which they are stable and precipitation at the lower limit. By this process progressively more saprolite would be brought into the profile where the differential solution and residual accumulation occur (Fig. 6). This is in fact precisely the process de Vletter (1955) believed to have been responsible for the

51

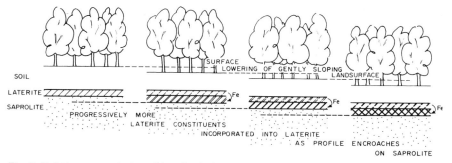

SOIL

LATERITE

SAPROLITE

SURFACE LOWERING OF GENTLY SLOPING LANDSURFACE

PROGRESSIVELY MORE LATERITE CONSTITUENTS INCORPORATED INTO LATERITE AS PROFILE ENCROACHES ON SAPROLITE

Fig. 6. *Relative accumulation of laterite during gentle surface-reduction.*
A process of continuous solution at the upper limit of the laterite horizon and deposition at the lower limit brings about progressive enrichment of the laterite horizon.

development of the Cuban laterites which he described in an extremely detailed and interesting account of their genesis (cf. also Nagell, 1962).

The re-solution of laterites under forests on eroded laterite surfaces is invalid evidence against the thesis that laterite can develop under such vegetation. Furthermore, such re-solution even provides a credible mechanism for the *accumulation* of a residuum, in association with a down-wasting and forested topography.

In conclusion, three lines of evidence contribute to the question of the vegetation suitable for laterite genesis. First is the circumstantial evidence of the often-seen association of laterite and grassland, which favours the thesis that laterite is a precipitate forming at the upper limits of a seasonally fluctuating water table. The edaphic nature of many of these grass climaxes renders this evidence of doubtful value. In many cases the grass exists *because* of the laterite. The not infrequent reports of laterite under existing forest or demonstrably recently existing forest (e.g. Chaplin and McFarlane, 1969) support the thesis that forest and laterite genesis are not incompatible. Second, detailed studies of the effects of vegetation on desilicification and iron stability suggest that well-vegetated conditions are not inherently unfavourable to the development of laterite. Third, the destructive role of forests on truncated or incized and unstable profiles is unsound evidence against the compatibility of forest vegetation and laterite genesis. The re-solution of precipitates in a downward-moving profile provides a means of accumulation of a residuum. It seems therefore that the entrenched belief that laterite genesis is exclusive to, or at least favoured by, a grassland vegetation is not warranted by the available evidence.

7

The Environment of Laterite — Laterite and the Profile

Laterite is too often discussed without reference to the most essential part of its environment: the profile in which it occurs. When considered in this proper context it becomes possible to eliminate some of the less well founded speculations concerning laterite genesis.

I. THE HORIZONS OVERLYING THE LATERITE

It is often difficult to establish whether these are original, modified, or derived, and therefore the degree of relationship with the laterite remains uncertain (Sivarajasingham et al., 1962). This is particularly true of the laterites on mesas. For example, over much of the Buganda Surface there are two soils. The younger appears to have developed from the older in response to local environmental changes, notably deforestation (McFarlane, 1968), but even of the older soil it is not possible to say with certainty to what extent it is "original". Certainly, material overlying laterite must be regarded with caution.

Red soils are commonly described in this situation (Humbert, 1948; Loughnan et al., 1962; Sivarajasingham et al., 1962). Dark humus-rich material is also reported to overlie laterite (Benza, 1836). Sometimes, as in Buganda (Pallister, 1954) the dark soils overlie the red in the central parts of the more extensive mesas. Elsewhere, only the red remains. Concretions may be found within the red soil, and these may become more closely packed with depth until they form a crust or indurated horizon (Sivarajasingham et al., 1962), the laterite proper. With such pedogenetic laterites, the precipitates evidently form within the soil where alternating conditions are provided and they are accumulated at the base of the soil. In a given static profile this process has obvious limits without a constant source of enrichment to the soil. Some thin pedogenetic laterites may possibly claim an entirely overhead source in this way, but in most laterites the accumulation is too great. As Trendall (1962) pointed out, to postulate an overlying source for the Buganada Surface laterite would presuppose the existence of an original 600 ft (183 m) of overlying material necessary to produce the concentrates. This is highly unlikely.

Pale-coloured leached horizons are also found overlying some laterites, for example the tropical podzol of Barshad and Rojaz-Cruz (1950). In this the ironstone layer, which qualifies as a laterite horizon by most criteria (Ch. 2), is apparently formed by leaching from the overlying A horizons. Once again, in a static profile such overhead contributions must be limited unless improbable thicknesses of formerly overlying material are postulated.

It seems clear that the large accumulations found in most laterite horizons cannot entirely originate in the overlying soil. It is nevertheless clear that in many cases, notably the pedogenetic laterites, the immediate source of enrichment is this overlying soil. Enrichment may be due to transport of iron from above by descending waters (e.g. Von Schellmann, 1964), or to the formation of concretions which settle to the base of the soil profile to form the laterite proper. It therefore becomes necessary to evoke some continuous enrichment of the overlying soil. For those laterites which occur on interfluves, lateral enrichment is precluded. Only one possible mechanism exists. The necessary replenishment of the overlying material is only possible if the profile is not static, but moving with a vertical component downwards. In this way there can be progressive incorporation of saprolite into the soil where the precipitates form or from where iron may be leached downwards. Fig. 7 outlines the main processes. By such a procedure large accumulations can be explained when the apparent source is the puzzling thin overlying soil horizons.

II.   THE LATERITE HORIZON

This may be of any thickness up to 200 ft (61 m) or more (e.g. Hays, 1967; Conniah and Hubble, 1960; Mikhaylov, 1964; Roy Chowdhury et al., 1965; etc.). Such thicknesses present a very serious

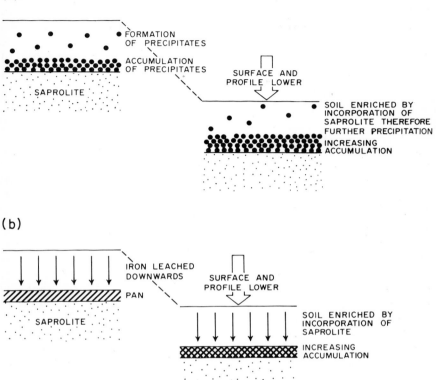

Fig. 7. *Possible modes of laterite formation from an overhead (pedogenetic) source.*
(a) Precipitates form in the soil and gravitate to the laterite horizon. (b) Iron is leached downwards from the soil into the laterite horizon.
In a static profile only small accumulations of laterite can develop from an entirely overhead source. Downward migration of the profile can allow larger accumulations to develop by continuous enrichment of the soil from unleached saprolite.

problem to the concept of laterite as a precipitate forming at the *upper limits* of the range of fluctuation of a static but seasonally oscillating water table. Pendleton and Sharasuvana (1946) stated that Campbell and Mohr "quite adequately explain how laterite develops in relation to the water table (and) why it is limited to a relatively thin horizon in the soil . . .". *The concept of laterite as a precipitate requires the laterite horizon to be thin*; but quite clearly in the majority of cases it is not (Goudie, 1973, p. 32). This forced Walther (1916) to suggest that laterite genesis occurred in a former climate, with extreme seasonal variations, which no longer exist today. Unwarranted as such a postulate appears, it seems that Walther at least recognized the problem of the scale of the laterite profile. Certainly it is quite impossible to reconcile such thicknesses with the seasonal oscillations of water tables in the essentially static profiles of planation surfaces.

III. THE HORIZONS BELOW THE LATERITE

Subsequent to Simpson's (1912) and Maclaren's (1906) descriptions of the profile underlying the laterite, Walther (1915) formalized the terms "mottled" and "pallid" zones.

Mottled horizons often underlie laterite (Benza, 1836; de Swardt and Trendall, 1970; Doyne and Watson, 1933; du Preez, 1954; Moss, 1965; Hays, 1967; Stephens, 1961). Fisher (1958) recorded 10—20 ft (3—6 m) of mottled zone, Humbert (1948) 3—25 ft (1—7.6 m), Loughnan *et al.* (1962) 7—29 ft (2.1—8.8 m), Mabbutt (1961), Hays (1967) and Walther (1916) up to 30 ft (9.1 m). According to theories of laterite as a precipitate, this mottled zone, up to 30 ft (9.1 m) thick, is the zone between the pallid zone and the laterite in which incomplete precipitation occurs. Again the problem of the scale of the assumed water table fluctuation arises.

The junction between laterite and mottled zone may be either sharp or gradational and the mottled zone grades downwards into the underlying material. There seems to be some confusion of laterite and mottled zone. For example Humbert (1948) described "iron crust" underlain by a "zone of concretions", while many laterites are described as being composed of mottles and concretions. Since the relationship of laterite to mottled zone is often gradational some of the recorded occurrences of very thick laterite may refer to both horizons. This does not affect the problem of the scale of the profile, for whether the material is called laterite or mottled zone the water table is required nevertheless to fluctuate through it.

The retention of rock structures has sometimes been used as a criterion to distinguish between laterite and mottled zone. Structures are usually preserved in the mottled zone but may be absent from the laterite (Doyne and Watson, 1933), indicating that the mottled zone is *in situ* and the laterite in some way derived (Benza, 1836; Blanford, 1859). This appears to support the concept of laterite as a residuum of weathering rather than a precipitate. However, rock structures are also recorded in the laterite. Some such reports may be attributed to a confusion of laterite and mottled zone, but not all.

56

The retention of rock structures in the laterite has been used to support the concept of it as a precipitate. However, the validity of this argument is doubtful for, as outlined elsewhere (pp. 51—52) a residuum need not be entirely mechanical. A residual precipitate such as that of de Vletter (1955) accumulates by a continuous process of solution at the top of the zone of precipitate stability and deposition at the bottom, thus leaving the zone of accumulation mechanically undisturbed. Nagell (1962) has also described the development of the residual Serra do Navio manganese ore laterites as a process of repeated solution and deposition. In short the retention of rock structures in the laterite as well as in the mottled zone is quite compatible with both the residual and precipitation models of laterite formation.

Although reputedly part of the typical profile, mottled zones are frequently absent (Blanford, 1859; de Vletter, 1955; Frasché, 1941; Hardy and Follett-Smith, 1931; Holmes, 1914; Lake, 1933; Ollier, 1959; Sivarajasingham et al., 1962). Sometimes the mottled zone is replaced by a yellow horizon (Frasché, 1941; Hardy and Follett-Smith, 1931) and sometimes it is underlain by this (Moss, 1965).

Below the mottled zone of the typical profile lies the pallid zone. White clays were recorded below the laterite and mottled zones in some of the early descriptions of the profile (Benza, 1836; Blanford, 1859) but it is to Simpson (1912) and Maclaren (1906) that recognition of the existence of this horizon is credited (Prescott and Pendleton, 1952). *This zone is by definition pale-coloured due to leaching of iron* (see Plate 12), and the coexistence of an iron-depleted and an iron-enriched horizon is the major piece of circumstantial evidence which gave rise to the theories of laterite as a precipitate. It was believed that the iron accumulation in the laterite originated in the pallid zone.

It is disappointing to have to reject such a convenient explanation, but the following problems facing it appear to be insurmountable.

(a) There is again the enormous scale of the assumed water table fluctuation. Fisher (1958) recorded pallid zones of up to 40 ft (12.1 m) thick, Blanford (1859) 44.5 ft (13.5 m) thick, Walther (1916) 5—15 m (16.4—49.2 ft), Mulcahy (1960) 60—80 ft (18.2—24.3 m), Hays (1967) 30 m (98.4 ft), Jessup (1960) 60 ft (18.2 m) to over 200 ft (61 m), and de Swardt and Trendall (1970) over 200 ft (61 m). It is quite inconceivable that water tables fluctuate through such a range that such thick pallid zones could occur in the lower part.

(b) It appears that the depletion of even the deepest pallid zones is nevertheless inadequate to account for the enrichment of the laterite (Trendall, 1962).

(c) There is no known mechanism for such upward movement of iron. Capillary action must be discredited (p. 93). If water table oscillations are invoked even this cannot explain the actual upward movement of enriched waters. As Sivarajasingham *et al.* (1962) noted, the development of water supplies in Hawaii is based on the principle that the rise of the water table during wet seasons is by superimposition of relatively fresh water which "floats" on the underlying solutions.

(d) The pallid zone is often absent even below *in situ* laterite (Mulcahy, 1961; de Swardt, 1964; Nye, 1955; Maud, 1965; McFarlane, 1969). It is therefore difficult to visualize laterite and pallid zone development as essentially *synchronous* and *complementary* processes.

(e) The supposed leaching of many of the so-called pallid zones has not in fact been proved. For example, the 200 ft (61 m) pallid zone described by de Swardt and Trendall (1970) was not analysed to establish that it is in fact depleted (A. M. J. de Swardt, personal communication). Pallid zones may be characterized by 2 : 1 clay minerals which are white in colour yet retain the bulk of the iron content of the parent minerals from which they are derived (Nye, 1955). Two deep "pallid zones" were analysed by the writer in Uganda, one below the low-level laterite of Busia, and one below the high-level laterite on Nsamizi Hill (see Plate 13). Although in the descriptive sense undoubtedly pallid, neither were in fact found to be depleted (McFarlane, 1969, Chs 19 and 23).

(f) Where pallid zones and laterite do coexist, there is no consistently direct relationship between the degree of development of the pallid zone and the laterite. Mulcahy (1960) noted that the depth of the pallid zone decreases into the drier interior of Western Australia, but apparently noticed no comparable lesser development of the laterite. Loughnan *et al.* (1962) described a series of profiles reputedly formed by the upward migration of iron. In general, the profiles show *better* development of pallid zone where there is *less* enrichment of the overlying horizons, and the only profile they describe as a true laterite profile shows virtually no pallid zone. Thus, the relationship between laterite and pallid zone development appears in this case to be inverse. Blanford (1859) also noted this relationship, that is, "where laterite is thin or only slightly ferruginous the clay appears to be nearly or entirely destitute of iron beneath". Such observations are clearly difficult to reconcile with the concept of laterite and pallid zones as synchronous and complementary developments.

(g) There are numerous occurrences of *in situ* laterites or bauxites resting directly on fresh rock (Fisher, 1958; Holmes, 1914; Lake, 1933; de Vletter, 1955; Frasché, 1941). Some cases may be explicable by lateral movement and precipitation of iron-rich solutions from adjacent higher lying weathered rocks, but this cannot explain such laterites on interfluves. In many cases there is in fact a profile between laterite and rock, but it is very narrow. For example, Frasché (1941) noted that the contact is sharp at first sight but on closer examination the underlying serpentine grades upwards into a narrow greenish yellow transition zone. Similarly, de Vletter (1955) observed that what appears to be a sharp division between serpentine and laterite is in fact a gradation, often only a few inches wide. In this case the laterite is up to 80 ft (24.3 m) thick. The inadequacies of postulating complementary upper horizon enrichment and lower horizon depletion become even more obvious. Residual accumulation is clearly indicated.

In total then, the data becoming available on pallid zones does little to support the post-Campbell concept of laterite as a precipitate. If the depletion of the pallid zone was not upwards into the laterite, then it must have been downward. No mechanism is known for its upward translocation but downward loss by leaching is a well known process. Loss of bases occurs readily and in tropical environments silica is also lost. Iron is not entirely stable and losses certainly occur in the weathering profile (Trendall, 1962). The inverse degree of development of some laterites and pallid zones such as those of Loughnan *et al.* (1962) may therefore be compatible with the residual development of a material only partially stable in the profile. In a profile in which iron is completely stable and resistant to the process of differential solution during weathering, there would be no loss from the profile and residual accumulation would occur in the upper horizons (Fig. 8a); where iron is less stable and is in part lost during the differential solution there would be a partially depleted zone underlying a lesser surface accumulation (Fig. 8b); if iron is completely unstable, truly depleted zones would occur and there could be no residual accumulation (Fig. 8c) in the surface horizons.

Partial downward loss of iron during laterite formation thus

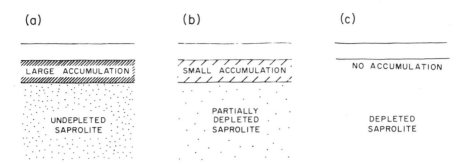

Fig. 8. *Inverse relationship between the amount of accumulation in the laterite and the degree of depletion of the saprolite.*
This sometimes observed relationship may be explained by considering the laterite as a relative accumulation of constituents over variously depleted saprolite.

provides one mechanism whereby a leached zone may develop contemporaneously with the laterite horizon.

Clearly such a process cannot explain the enormous concentrations which overlie entirely depleted pallid zones such as are found blanketing interfluves in many parts of the tropics; and it is with these that the geomorphologist is primarily concerned. A further piece of circumstantial evidence provides one plausible answer to the problem. Pallid zones are unique to laterites, but do not occur under all laterites. It has yet to be demonstrated that well developed pallid zones underlie *unincized* laterites. They have only been found below the high-level laterites, characteristically preserved on mesas standing hundreds of feet above the surrounding lowlands. If this generalization is valid, i.e. that pallid zones are exclusive to incized profiles, then it must be concluded that *pallid zone development is consequent upon incision of the laterite profile,* that is, it is a postincision modification of the profile. Many recent observations support this suggestion. Pallid zones have been observed to be currently subject to through-flushing. Mabutt (1961) noted the occurrence of seepage through the pallid zone which caused "cavernous weathering" and Stephens (1961) observed water flowing down the face of a cutting in the pallid zone in Buchanan's type locality. Alexander (Sivarajasingham *et al.*, 1962) believed that some of the pallid zones of Australia might still be forming today. In Buganda, those high-level laterites which were formerly forested and permeable are markedly deformed by subsidence so that in detail they are very reminiscent of karst topography (Fig. 9, and Plates 5 and 7). Reports of the formation of such pseudo-karst topography

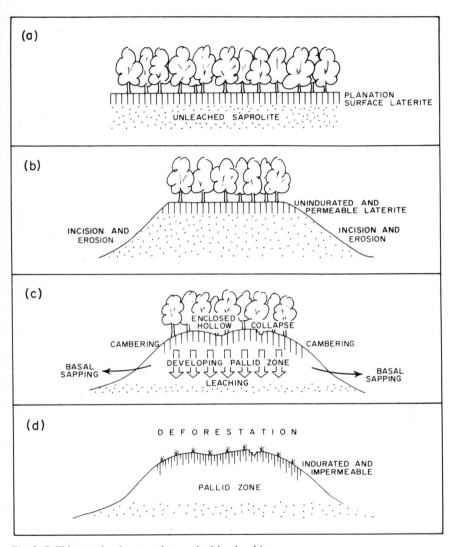

Fig. 9. *Pallid zone development by postincision leaching.*
(a) Residual laterite develops. (b) The laterite is incized. It remains permeable while the original soil cover and forest vegetation remain. (c) Leaching through the carapace depletes the underlying saprolite and pseudo-karst surface features develop, e.g. marginal cambering of carapace, enclosed hollows and collapse features. (d) Deforestation leads to exposure and induration of the laterite which loses its permeability. Leaching and pallid zone formation cease.

PLATE SECTION

Legends to Plates

PLATE 1. *Spaced pisolithic laterite.*
In this immature groundwater laterite, which occurs in Buganda, the generally wine-coloured pisoliths are unbanded and occur spaced within a white matrix of kaolin and quartz. The laterite is easily broken down to a rubble of pisoliths. (Pin for scale.)

PLATE 2. *Packed pisolithic laterite.*
This laterite is the detritus or remanié from the breakdown of the spaced pisolithic laterite shown in Plate 1. The pisoliths are identical but have an outer, yellow-coloured, cutane (C). They rest one against the other, with a red, soil-like material commonly filling the voids between them.

PLATE 3. *Pedogenetic laterite on fresh rock.*
Pedogenetic laterite commonly forms where fresh rock nears the surface. This produces locally extreme conditions of wetting and drying, as on the deforested and indurated surfaces of the mesas underlain by ancient groundwater laterite (Plate 4). This type of laterite therefore occurs in a variety of positions in the Buganda topography and is not of cyclic significance. [The "panga" handle is 7 in (18 cm) long].

PLATE 4. *A Buganda Surface laterite profile.*
A very thin topsoil with a grass cover overlies a horizon of pedogenetic laterite (A), about 1 ft thick. This rests on massive vermiform laterite (groundwater), the upper horizons of which are indurated and fractured (B), and the lower horizons soft and crumbly (C). At a depth of about 12 ft (3.6 m) (out of the picture) the vermiform laterite overlies spaced pisolithic laterite with a sharp junction. The veneer of pedogenetic laterite encloses artefacts. Its development and the hardening of the old groundwater laterite are the relatively recent results of deforestation of the old laterite surface.

PLATE 5. *Surface solution of the laterite carapace.*
A grid of fractures in the laterite carapace, at the margin of a subsidence hollow, has been picked out by surface solution to form microrelief features very similar to the clints and grykes of limestone surfaces.

PLATE 6. *Forest clearance on the Buganda Surface.*
Forest clearance in the centre of one of the larger Buganda Surface mesas exposes the massive vermiform laterite (foreground) from beneath a thin protective soil cover.

PLATE 7. *The high relief of the "Buganda Surface".*
The grass-covered sheet of laterite in the foreground slopes continuously upward behind the camera to reach a comparable altitude to the mesa seen on the skyline, that is, about 250 ft (76 m) above the camera position. The forest in the middle distance marks the carapace edge. The irregular line of fresh grass running diagonally away from the camera marks an ephemeral stream, with small collapse holes along its course.

PLATE 8. *Massive vermiform laterite.*
This continuous phase variety of laterite which occurs in Uganda is characterized by pipes containing kaolin and quartz. At depth it is soft and fragile. When exposed on the surface it hardens and darkens, the pipe infill is washed out and the pipes become lined with concentric layers of dark brown goethite. Bar represents 1 in (2.5 cm).

PLATE 9. *The formation of a packed pisolithic laterite.*
The yellow cutanes are conspicuous on this rubble of pisoliths eroded from an exposure of spaced pisolithic laterite on the Buganda Surface. A fragment of spaced pisolithic laterite, still intact, can be seen in the upper right hand quarter of the photo.

PLATE 10. *An exposure of low-level spaced pisolithic laterite.*
Murrum quarries expose the spreads of immature spaced pisolithic laterites which occur within the gently rolling interfluves of the low-level surfaces in Uganda. Such groundwater pisoliths are no longer forming; the water table has lowered, leaving them "high and dry". With progressive down-wasting of the interfluve they will become incorporated into the residual mantle of pisoliths which drapes the interfluve surface.

PLATE 11. *A sample of partly altered packed pisolithic laterite.*
The pisoliths in this core sample are a detritus from an original spaced pisolithic laterite. The outer yellowish cutanes are clearly visible. The subsequent grouping of the pisoliths and the development of nascent pipe structures between the groups can also be seen. (Match for scale.)

PLATE 12. *A pallid zone exposure near Kampala.*
Outcrops of leached saprolite are found 200 ft (61 m) or more below the tops of laterite-capped mesas. The material is very permeable, as seen by the small potholes.

PLATE 13. *A core sample from below the "high-level laterite" in Buganda.*
The core, drilled through a "high-level laterite" carapace, passed through more than 200 ft (61 m) of material which was pallid in the descriptive sense, but was undepleted of iron, and more correctly described as saprolite. Since the carapace was of detrital packed pisolithic laterite, the level of the *in situ* laterite was above the present summit of the hill and this shows that the depth of weathering below the *in situ* Buganda Surface laterite was well in excess of 200 ft. Such deep weathering indicates an abundance of moisture, rather than the more limited moisture available where marked seasons occur. (Match for scale.)

PLATE 14. *A pedogenetic laterite overlying a residual sheet of packed pisolithic laterite.*
This profile was exposed in a murrum pit cut into the mantle on the crest of a gently rolling interfluve in the Kyoga lowlands (Wayland's PIII).

PLATE 15. *Naminya pediment — an example of relief inversion from Buganda.*
The grass covered pediment (A), underlain by laterite, is becoming isolated from the mesa (B) by the development of a small valley (C) between it and the main hill. The former pediment is beginning to emerge as a small isolated mesa. To the right, outside the picture, the pediment is still continuous with the main hill. Its isolation and the process of relief inversion is not yet complete.

PLATE 16. *The Buganda Surface in the type area, Kampala.*
Here, because the mesas are very small each appears very flat; strongly suggesting that they formed part of a well-developed planation surface. The flat summits fail, however, to fall within a narrow altitude range. Scale = 1 : 40,000.

PLATE 17. *The Buganda Surface in Kyaggwe.*
Here, the larger mesas show more clearly the form of the "high-level laterite surface". The northern ends of the two wings of the hill are altitudinally about 250 ft (76 m) lower than the southern ends. The higher areas in the south are underlain by massive vermiform laterite and are typically dimpled with small depressions. The two wings of the hill are separated by a small pisolith-strewn col where the underlying spaced pisolithic laterite is exposed. Where the two wings of the hill narrow northwards, spaced pisolithic laterite outcrops again, with packed pisolithic laterite immediately downslope. Further to the north of this the widening portions are underlain by a second younger, vermiform laterite, here 150—200 ft (46—61 m) lower than the first vermiform laterite on top of the hill. The lowest most northerly extremities of the hill are capped by packed pisolithic laterite, the detritus of spaced pisolithic laterite underlying this second vermiform laterite. Despite the continuity of the laterite sheet on this mesa it is not a chronological entity. Scale = 1 : 40,000.

PLATE 1

PLATE 2

PLATE 3

PLATE 4

PLATE 5

PLATE 6

PLATE 7

PLATE 8

PLATE 9

PLATE 10

PLATE 11

PLATE 12

PLATE 13

PLATE 14

PLATE 15

PLATE 16

PLATE 17

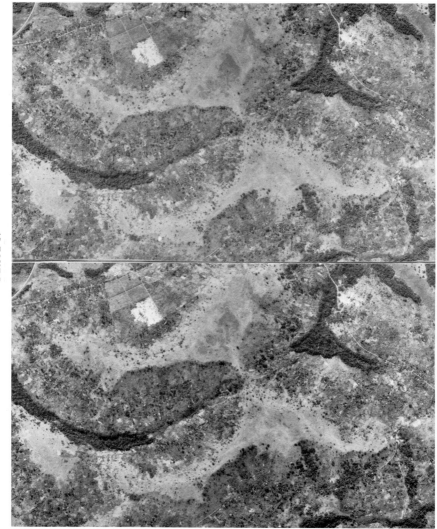

on laterite surfaces are becoming increasingly frequent (Goudie, 1973, p. 46; Thomas, 1968). Through-flushing is clearly indicated. In Buganda the only high-level mesas which are undeformed are capped by impermeable detrital laterite and a core over 200 ft (61 m) deep through one such carapace showed what appeared to be a pallid zone but this when analysed showed no depletion of iron (McFarlane, 1969). Furthermore, pallid zones have not been found beneath the low-level *in situ* laterites of Buganda. The development of these laterites can be explained as entirely residual and, since the high-level laterites are essentially similar, an entirely residual origin, without contribution from the pallid zone, is also postulated for their development (McFarlane, 1971).

In short, the pallid zone, at first sight providing an answer to the question of the provenance of the concentrates, creates more problems than it solves for those who followed Campbell in believing laterite to be a precipitate. On closer inspection it has become in itself an end which must be explained, rather than a means of explaining the laterite formation.

Summary

Study of the relationship of laterite to the adjacent horizons casts considerable light on the geomorphic environment of laterite genesis. The overlying soil cannot provide the concentrates *if the profile is stable*; a downward migrating profile must be evoked if lateral contributions cannot be found. Even where underlying iron-depleted horizons occur, there is no way by which the loss of iron from them can contribute to the overlying accumulations in the laterite. Again, an overhead source of enrichment must be sought. The profile therefore presents evidence to strongly suggest that we must look for a model of laterite formation which does not limit its development to the stable profile of a planation surface. Residual accumulation *during landsurface reduction* is clearly implied. The enormous scale of the typical profile argues strongly against the *monogenetic* development of the various horizons, and in particular against the synchronous and complementary development of the lower depleted horizons and the upper enriched horizons. The typical profile of laterite and mottled zone overlying a depleted pallid zone, which led to the belief that laterite is essentially a precipitate, in fact supports more readily the concept of laterite as a residuum.

8

Laterite Structures

Laterite occurs in various morphological forms. That most commonly described is the "discrete concretion" variety, in which the bodies are often loosely described as nodules, concretions, pellets or shot. More specifically these bodies have been classified into *pisoliths* and *ooliths*, the former being over and the latter under 2 mm diameter as suggested by Twenhofel (du Preez, 1954). There is some confusion here. Although many authors use the terms pisolith and oolith (e.g. Johnson, 1959; M. J. Mulcahy, personal communication; Ollier, 1959; Pullan, 1967; Maignien, 1966) some use the terms *pisolite* and *oolite* to describe the individual bodies (e.g. de Preez, 1954; Hanlon, 1945; Alexander and Cady, 1962; Faniran, 1970, 1971; Campbell, 1917), when these terms should be reserved for the *rocks* composed respectively of pisoliths and ooliths (Stamp, 1961; Pullan, 1967). That is, a rock-like mass of pisoliths is a pisolite and such a mass of ooliths is an oolite. Logically the adjective from pisolith is *pisolithic* as used by Brosh (1970), Maud (1965), Moss (1965), Johnson (1959), McFarlane (1971, 1973) and Sombroek (1966). A *pisolithic laterite* is one composed of pisoliths, not necessarily cemented into a mass or pisolite. A laterite composed of a cemented rock-like mass of pisoliths is a pisolite or *pisolitic laterite*. Confusion arises when authors use the term pisolitic as the adjective to describe all laterites containing pisoliths, whether or not the laterite is the rock-like cemented variety (e.g. Maignien, 1966; du Preez, 1954; Campbell, 1917; H. Osmaston, personal communication). Not all laterites containing pisoliths are rock-like and pisolitic. In fact two broad groups are recognized: those in which the pisoliths are cemented together and those in which they occur spaced within an earthy matrix (see Plates 1 and 2). For geomorphology this terminological problem is no academic exercise. It is very important to distinguish between the two types. As far as groundwater laterites are concerned, pisolitic laterite is a remanié or detritus, while pisolithic is *in situ*. In Uganda, mesas capped by the sheets of closely packed pisoliths are well defined, with a sharp free face or scarp edge, while the mesas capped by the spaced variety are recognizable by their poorly defined form and absence of free face since this laterite falls apart on exposure. In attempts to reconstruct former landsurfaces in Uganda it was found that the well-defined mesas capped by cemented pisoliths were misleading because this is a detrital laterite. Those capped by spaced pisolithic laterite were meaningful (Fig. 10), representing the lower parts of an *in situ* profile.

It is imperative for geomorphologists to distinguish between these two varieties of pisolithic laterite. To confine both the detrital and *in situ* varieties under one term, most popularly "pisolitic laterite", is clearly unsatisfactory. Since there are as yet no established terms to distinguish them, the need for working terms in Uganda led to the use of "packed pisolithic laterite" and "spaced pisolithic laterite" (McFarlane, 1971, 1973) which proved serviceable.

In other respects also, the terms pisolithic and oolithic create problems. Since the distinction is based arbitrarily on size it is hardly meaningful (Pullan, 1967) and is sometimes ignored. For example, Hanlon (1945) says "the pisolites [sic] (i.e. pisoliths) vary from microscopic dimensions to more than an inch in diameter". Further

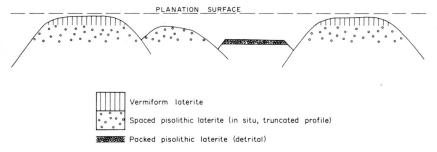

Fig. 10. *The relationship between the nature of the laterite on mesas and the position of the former landsurface.*
The tops of mesas capped by spaced pisolithic (*in situ*) laterite truncate the profile and therefore fall short of the original height of the laterite surface, marked by the remaining vermiform laterite. Those capped by packed pisolithic laterite, the detritus from this erosion, occur at a variety of lower altitudes.

difficulties arise since many pisolithic laterites contain both ooliths and pisoliths (Pullan, 1967; McFarlane, 1969, 1971, in press). In Uganda, the spaced pisolithic laterite profile, some 60 ft (18.2 m) deep was found to alter with depth, ooliths being present in the upper horizons as well as pisoliths, while pisoliths predominate a little lower, and in the lower parts large concretions coexist with the pisoliths (Fig. 11). Since pisoliths are present in all horizons, the gross descriptive term used was "spaced pisolithic laterite". In reconstruction of the former landsurfaces this distribution pattern in depth allowed a rough assessment to be made of the extent of the

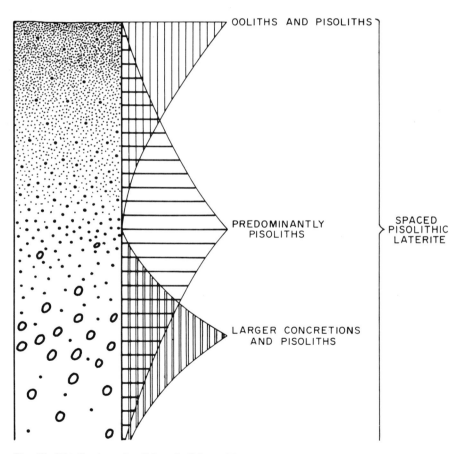

Fig. 11. *Distribution of ooliths, pisoliths and larger concretions within the spaced pisolithic laterite profile.*

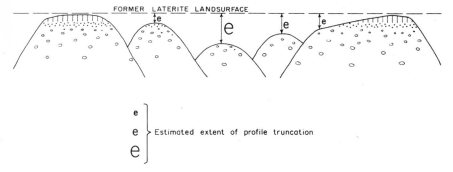

Fig. 12. *Truncation of the spaced pisolithic laterite profile.*
Estimates of the extent of truncation can be made from the proportions of ooliths, pisoliths
and larger concretions in the exposed laterite.

truncation of the profile exposed on the surfaces of mesas
(McFarlane, 1973) (Fig. 12).

A further problem with the terms pisolith and oolith is the lack of
agreement about whether or not a concentric laminar structure is
diagnostic (Rice, 1957; American Geological Institute, 1962; Page,
1859; Challinor, 1961; Schieferdecker, 1959; Campbell, 1917;
Pullan, 1967). Many authors use pisolith simply to mean "resembling
a pea" and they specify in the text the nature of the internal
structure, a practice adopted by the present writer. Pullan (1967)
suggested that the unbanded variety be *nodules* and the banded
variety be ooliths and pisoliths, but the term nodule is already
popularly used to indicate a rounded lump of varied origin —
rounded or rolled fragments of any of the other kinds of laterite as
well as other precipitates cemented into a group.

There is more agreement about the formation of such bodies.
They form by centripetal enrichment in the zone of fluctuating
water table (Pullan, 1967). Mottles are poorer developments of the
structure, which reaches its perfection in the pisolithic or oolithic
form. Unbanded pisoliths are believed to form where the substance is
pure, while the presence of other substances, colloid or crystalloid,
which are also precipitated, causes a banded structure (Rao, 1928;
Hanlon, 1945). Pisoliths which form in the soil, the pedogenetic
pisoliths, are often associated with manganese precipitates and are
irregular in shape and size (Cornwall, 1958; Beater, 1940; Doyne and
Watson, 1933; McFarlane, 1971). They enclose soil-like material and
often have a dark outer layer of manganese. The "shot" or truly
pea-shaped pisoliths, groundwater precipitates which form in

saprolite, are often very regular and well-rounded and enclose saprolite-like material similar to the saprolite in which they form. In Uganda the *in situ* groundwater pisoliths, that is those spaced within a saprolite matrix, have no outer banding, while those which have been eroded from their position of formation had developed an outer cutane of yellow material (see Plates 1, 2 and 9). Such yellow cutanes occur only where the groundwater pisolithic laterites have been reworked, that is where they are a remanié or detritus of the *in situ* variety. Pullan (1967) also observed that this shiny yellow patina is characteristic of transported pisoliths or nodules.

In detail, pisoliths may be cracked, possibly due to shrinkage from the gel state to the solid state (Lindgren, 1925). Aluminous pisoliths also occur (Van Bemmelen, 1941; Hardy, 1931). Pisoliths have the ability to seal within them and preserve relatively coarser (Moss, 1965) or unweathered material (du Preez, 1949; Alexander and Cady, 1962; Sivarajasingham *et al.*, 1962; Mulcahy, 1961; Mulcahy and Hingstone, 1961; McFarlane, 1969).

This ability of groundwater pisoliths to "fossilize" internally the state of weathering of the saprolite in which they form has provided a means of explaining the very deep profiles which exist in Uganda (McFarlane, in press). There some 60 ft (18.2 m) of spaced pisolithic laterite underlies the planation surface laterite (which has a completely different vermiform structure). An examination of the pisoliths at the top of such a profile showed that the saprolite they contain is less weathered than the saprolite between them. Presumably the matrix continued to weather after the pisoliths formed, and the weathering front advanced in depth. At the base of the profile the general state of weathering is comparable to that indicated by the pisoliths in the top of the profile, which suggests that the higher pisoliths formed when the matrix was in a similar state to that now found 60 ft (18.2 m) lower. The sequence of events would appear to be as follows. Weathering proceeded from the top of the profile downwards. The pisoliths at the top of the profile formed when the state of weathering was much as it is now at 60 ft (18.2 m). Subsequent to their formation weathering proceeded further, that is, the matrix weathered further and the weathering front advanced. The formation of the pisoliths in the lowest position must therefore postdate the formation of the pisoliths at the top of the profile, and thus pisolith formation can be visualized as occurring at progressively lower levels. The formation of pisoliths is associated with an oscillating water table and thus we can visualize a gently lowering oscillating water table, presumably a water table lowering in response

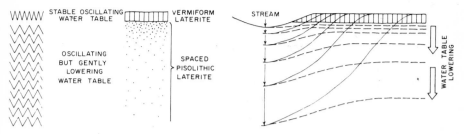

Fig. 13. *Water table movements associated with the formation of the high-level laterite profiles in Uganda.*
Mature massive vermiform laterite, the residual end-product of one erosion cycle, indicates a period in which the water table is stable but oscillating. Upon rejuvenation and incision of the surface, iron in the undepleted saprolite is segregated into immature spaced pisolithic laterite as the oscillating water table gently lowers. With increasing depth the spaced pisolithic laterite becomes more poorly developed, and fades out at about 60 ft (18.2 m).

to renewed landsurface incision after a stable period associated with the formation of the overlying planation surface laterite (Fig. 13). By this process it becomes possible to understand the extraordinarily deep profiles without recourse to unlikely water table movements (p. 56).

The ability of pisoliths to preserve relatively fresh material within them has led to the conclusion that they form early in the process of lateritization (Alexander and Cady, 1962), that is, *the concretion stage is a definite step in the genesis of laterite* (Humber, 1948). This suggested immaturity of the pisolithic form of laterite accords with the observation, discussed elsewhere (Ch. 4) that this structure often occurs in association with relatively high relief. It is also interesting that the pisolithic structure of some calcretes has been suggested to belong to immature varieties (Netterberg, 1966; McFarlane, 1976). Pisolithic laterite, an immature form of laterite, appears to be associated with what might be termed "immature planation surfaces". Maud (1965) noted that concretionary laterites characterize the less well developed post-Caenozoic surfaces while the Caenozoic surface bears a vesicular laterite. In Uganda the rolling interfluves in the Kyoga Basin are characterized by pisolithic laterites (Fig. 14b and Plates 10 and 14). Only in the lower parts of the catenas are vermiform structures developed. The large vertical spread of pisoliths within such interfluves can again be associated with a lowering water table. Such landforms are by no means flat and are actively down-wasting. The sheets of packed pisolithic laterite have accumulated as a remanié in the base of the soil by a process of soil encroachment on pisolith-bearing saprolite. With this surface reduc-

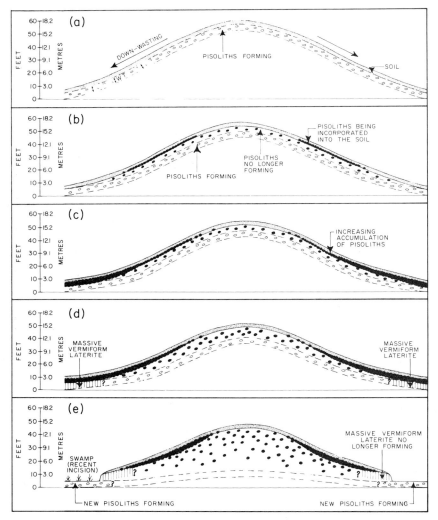

Fig. 14. *Immature, pisolithic laterite development in the Kyoga lowlands.*
(a) Pisoliths form within the range of water-table oscillation, in association with a gently rolling low relief landsurface. (b) Down-wasting lowers the groundsurface and water table, thus lowering the locus of pisolith formation. Pisoliths become incorporated into the base of the soil as this encroaches on the underlying saprolite. (c) Further landsurface reduction is accompanied by the development of an increasingly thick sheet of residual pisoliths. (d) In low catenary positions massive vermiform laterite develops by alteration and hydration of the sheet. (e) Incision terminates the process of laterite development before the landsurface is sufficiently reduced for mature laterite to be extensively developed. (For clarity, slopes have been exaggerated in this diagram.)

tion the water table also lowers, and with it the locus of current pisolith formation (Fig. 14a,b,c).

It is perhaps no coincidence that pisolithic laterites are the most frequently described. If they are immature forms of laterite this is to be expected, for imperfectly developed planation surfaces or partial surfaces invariably occur more frequently than perfect planation surfaces.

The degree of hydration of the contents of such pisoliths also accords with this concept of their early formation in association with the more aerating conditions of a rolling topography and non-static water table, for they characteristically contain a high proportion of haematite, as well as goethite, in contrast to the more "limonitic" content of the massive laterites or planosol developments.

The distribution of pisoliths within the profile, already touched upon elsewhere (Ch. 7), may be summarized thus. If the pisoliths form in the overlying soil, they increase in frequency towards the base (e.g. Ollier, 1959; Doyne and Watson, 1933; Sivarajasingham *et al.*, 1962), where they may form a packed sheet or overlie a more massive laterite (Fig. 15a). It has been suggested that this pattern of distribution reflects the suitability of the different parts of the profile to precipitate formation; that is, where the pisoliths are thinly scattered the suitability for pisolith formation is less than it is lower in the profile where they are densely packed. This may be so, but since the pisoliths themselves are as well developed in the upper as in the lower horizons, it seems more likely that this distribution pattern is the result of the natural migration of such heavy bodies towards the base of the soil, by a process of soil creep or by the activity of soil fauna e.g. worms and termites.

If pisoliths form in the saprolite, the distribution pattern is the reverse (Fig. 15); they become less well defined, less frequent and more irregularly shaped with depth until they grade into mottles (du Bois and Jeffery, 1955; Mulcahy and Hingstone, 1961; Doyne and Watson, 1933, McFarlane, 1969). Since laterite is essentially a near-surface formation, this distribution pattern may indeed reflect the progressively less favourable environment with depth. An overlying sheet of closely packed pisoliths often occurs in the base of the soil proper (Fig. 15b). Here the groundwater pisoliths lie juxtaposed in a matrix of soil. Such sheets are essentially detrital, the remanié of reworking of the spaced pisolithic laterite. Reworking to provide sheets of loose pisoliths was recognized early (Woolnough, 1918). The murram of East Africa and the valley side gravels described by Mulcahy and Hingstone (1961) and Stephens (1961) are examples of

(a)

PEDOGENETIC
PISOLITHS

(b)

GROUNDWATER
PISOLITHS

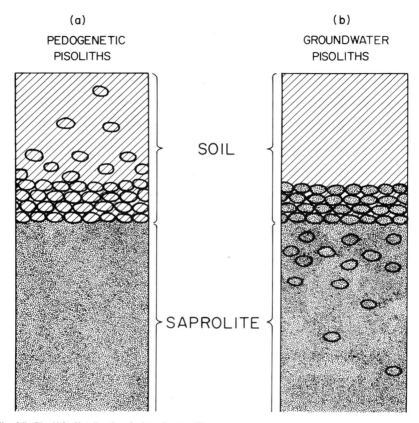

SOIL

SAPROLITE

Fig. 15. *Pisolith distribution in laterite profiles.*
(a) Pedogenetic pisoliths form in the soil and increase in frequency towards the base. (b) Groundwater pisoliths form in the underlying saprolite and increase in frequency upwards towards the base of the soil.

this. This remanié or detritus may become cemented together into a sheet (Benza, 1836). Such cemented sheets of packed pisoliths occur commonly in Uganda, and their hardness is responsible for the well defined form of mesas capped by this kind of laterite.

The other main type of laterite structure is the continuous phase variety, in which the indurated elements form a continuous coherent skeleton. The terms used to describe the structures include *vermicular* (Hanlon, 1945; Stephens, 1961; Russell, 1962; de Preez, 1949), *vesicular* (Brosh, 1970; Russell, 1962; Moss, 1965; Maud, 1965; Humbert, 1948; Griffith, 1953; Sivarajasingham *et al.*, 1962; Stephens, 1961; Pendleton and Sharasuvana, 1946; de Swardt, 1964;

de Preez, 1949; Newbold, 1844; Maclaren, 1906), *cellular* (Brosh, 1970; de Preez, 1949; Falconer, 1911; Benza, 1836), *vermiform* (Johnson, 1959, 1960; Johnson and Williams, 1961; Hanlon, 1945; de Swardt and Trendall, 1970; Goudie, 1973) and *tubular* (Fisher, 1958; Teixeira, 1965; H. Osmaston, personal communication). The differences have never been clearly defined (Pullan, 1967). It is the *shape of the cavities* as opposed to the shape of the precipitates which is the main descriptive criterion. Cavities may be tube-shaped (Clark, 1838; Newbold, 1844; Johnson, 1959), pipe-shaped (Blanford, 1859; de Bois and Jeffery, 1955; Humber, 1948), kidney-shaped (Babington, 1821) or a great variety of irregular shapes. The terminological problem is clear from but a short review of the literature. For example, Teixeira (1965) describes tubular laterite. Johnson (1959) describes vermiform laterite as having tubular cavities, and a reader would therefore assume these laterites to have the same structures. Vermiform means "worm-shaped" (OED) and since Pullan (1967) describes vermicular as having sinuous worm-like tunnels, again these might be equated. Russell (1962) appeared to regard vermicular and vesicular as the same. Pullan (1967) considered that the shape of the cavities of cellular and vesicular laterite is the same (but in the former the earthy content has been removed). On the face of it, therefore, it seems that all of these terms describe the same structure! However, it is quite clear that the shape of such structures differs and often a given writer will differentiate between, for example, vermicular and cellular (Pullan, 1967) or vesicular and vermicular (Stephens, 1961). Pullan (1967) has attempted to give the structures serviceable definitions. He chose the terms "vermicular", "vesicular" and "cellular" to describe continuous phase precipitates. However, this terminology is questionable since he uses the terms to differentiate between *similar structures* which respectively contain their earthy infilling and those from which the material has been removed. The presence or absence of infilling seems a peripheral criterion and the present writer feels that the terms should be entirely structural, that is, they should describe the *shapes* of the cavities. In view of the widespread confusion and lack of definition, the terms adopted by the writer in Uganda were based on the OED definition of terms, which should be internationally intelligible. Vermiform was used in preference to vermicular for tube-like structures (see Plate 8), as vermiform is purely descriptive meaning "worm-shaped", while vermicular, meaning "worm-eaten (in appearance)" has genetic implications which would be unacceptable to many. Tubular is in fact equally good, but

75

vermiform is already in use in Uganda (Johnson, 1959, 1960; Johnson and Williams, 1961; de Swardt and Trendall, 1970). A vesicle is a small bladder, cell, bubble or hollow and cellular means having small single rooms, compartments or cavities. There seems to be little to choose between these two to describe laterites in which the cavities are irregularly shaped cells rather than tubes, but cellular seems marginally more appropriate.

This discussion of terms may seem irrelevant, but it is impossible to begin to discuss the regional or topographical significance of the various structures of the laterites when there is uncertainty about whether or not one author's vesicular laterite is the same as another's. The regular use of photographs in papers helps, but it is expensive, and problems still arise. For example, Pullan (1967) says "Plate XXIX of Russell (1961) should be described as vesicular cellular rather than vesicular". Certainly a great deal of confusion has arisen through the use of imprecise terminology. A conscious recognition of the problem will do much towards some rationalization in future literature.

There is less agreement about the formation of these structures than about the formation of the discrete concretion varieties. There is a small body of opinion which believes termites to be responsible for the tubular cavities (Russell, 1962; Osmaston, in McFarlane, 1971; Erhart, 1951, p. 806), but it is not generally supported (e.g. Griffith, 1953; de Swardt, 1964). The tubular cavities are more widely believed to be an integral part of laterite structures. Some believe the irregularly shaped cavities to be caused by the removal of kaolin and other earthy materials (Sivarajasingham et al., 1962; Griffith, 1953; Clark, 1838). Such explanations, however, do not actually explain why material should be differentially removed in such a way as to leave these structures. Some manner of movement of solutions is popularly believed to be responsible for the structures (Oldham, 1893; Hanlon, 1945; Blanford, 1859; Benza, 1836; Prescott and Pendleton, 1952; de Bois and Jeffery, 1955) a view to which the present writer subscribes. Much as the discrete concretions appear to be normal precipitation structures attributable to the movement of solutions, the present writer belives the tubular structures to be another form of normal precipitation structure.

Another proposed explanation is that they are formed by root penetration (Pendleton and Sharasuvana, 1946; Benza, 1836). Some tubular cavities are certainly caused by root penetration. Examples have been found in Uganda (McFarlane, 1969, and in press). There, root-formed cavities are nevertheless readily distinguishable from those which are an integral part of the laterite structure, and certainly the vermiform structures *on the whole* are not attributable to root penetration.

In general, truly vermiform or tube-like cavities seem to occur only in groundwater laterite. In Uganda it was found that within the vermiform laterite profile the arrangement of the tubes varied with depth, being predominantly vertical in the upper parts, with an anastomosing arrangement in the middle, and being predominantly horizontal towards the base.

Cellular laterites are usually pedogenetic and a platy structure, often with horizontal tubes, is commonly developed towards the base.

The separation of laterite structures into discrete concretion and continuous phase varieties is in some ways rather artificial, for these are merely the structural extremes which occur. Between these extremes lies a whole range of structures in which both vermiform and pisolithic structures are variously developed (Prescott and Pendleton, 1952). The regularity of the shape of the tubes and cavities varies and as the form becomes less regular, the discrete bodies become the more obvious structural feature. In laterites with irregular tube forms, mottles and irregular concretions often occur between them (Oertel, 1956; Oldham, 1893; Moss, 1965; Johnson, 1959, 1960; Stephens, 1961). When the structures are predominantly pisolithic, tubes as such are not usually present although cavities exist between the pisoliths, and on the exposed surface these may give a vesicular appearance to the laterite (du Bois and Jeffery, 1955).

The gradation between these various structures makes it highly unlikely that the mode of formation of vermiform and pisolithic structures is *entirely* different (Prescott and Pendleton, 1952), supporting the suggestion that these are simply variations on a precipitation theme. Alexander and Cady's (1962) studies of laterite microstructures, largely beyond the scope of this discussion, show the occurrence of microconcretions and the segregation of iron around pores and along channels. These are all precipitation features and there seems as little justification for postulating microtermites to explain those tiny channels as there is for postulating normal termites to explain the larger examples of channels.

If the pisolithic laterites are an immature form of laterite and the more massive, vermiform laterites are mature, then it seems reasonable that the latter develops from the former, as the landsurface with which they are associated approaches senility.

Detailed studies of detrital-packed pisolithic laterites in Uganda (McFarlane, 1969) have provided some information about the way vermiform structures develop as the detritus becomes altered in its new position. Vermiform laterites appear to develop both by alteration of detrital sheets of packed pisolithic laterite (see Plate 11) and by impregnation of the underlying saprolite with iron lost from the overlying, altering sheet. The chemistry of such alteration remains problematic (D. Yaalon, personal communication).

Can any generalizations be made about the occurrence of these various structures, generalizations which might be of value to the geomorphologist? Clearly the environment of formation differs. Some authors believe that lithology is a determining factor (Pullan, 1967; Goudie, 1973) while others believe this influence to be minor, other environmental factors dominating (McFarlane, 1969). Since both structures commonly occur in different horizons in a given laterite profile, non-lithological determinants must operate. *In a monogenetic profile* pisolithic laterites usually occur in the upper horizons, while vermiform structures underlie these (Ollier, 1959; Fisher, 1958; Blanford, 1859). A relatively less aerating environment of formation is indicated for the vermiform variety, which accords with the degree of hydration of their contents. On a larger scale this relationship is also to be found, for vermiform structures appear to be associated with lower relief than the pisolithic forms (pp. 36–37), or they occur as the lower members of a catena (McFarlane, 1969). These structures appear to belong to groundwater laterites which are truly planosol developments.

This generalization was found to be of practical application in Uganda where the mesas of the so-called Buganda Surface occur at a confusing variety of altitudes. When type of laterite was plotted, vermiform laterite was found to occur in two altitudinal groupings, and the laterites on mesas at intermediate altitudes were either detrital (packed pisolithic) or represented the lower parts of the truncated profiles (spaced pisolithic laterite). This is illustrated in Fig. 16 and Plate 17. Vermiform laterite occurs on the highest mesas on interfluves overlying quartzites which are weathered to some 200 ft (61 m) below the level of the laterite. These must represent the highest parts of the former landsurface and so it can be concluded that vermiform laterite occurred over the entire surface, not merely the lower parts of it, that is, it was a planation surface *par excellence.* The lower vermiform laterite, that of the succeeding surface, was very extensively developed but this surface had not entirely removed the residuals of the older planation surface, an

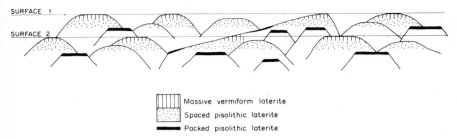

Fig. 16. *Diagrammatic representation of laterite on mesas in Kyaggwe.*
These represent two modified planation surfaces each marked by massive vermiform laterite.

estimated 2% of which survived. On the much younger surfaces, vermiform laterites are only poorly developed in the lowest parts of the catenas. Laterite structure here played a useful role in landscape analysis and a reconnaissance study of northern Zaire and the Central African Republic have shown that here too the laterite structure may be a useful parameter to study in this respect.

In conclusion it appears that the various structures reflect the local environment of formation of the laterite so that a genetic classification of laterite based on structure (McFarlane, in press) can provide a useful tool for the recognition of the nature of the imperfectly preserved landsurfaces on which these laterites occur.

9

The Chemical Constituents of Laterites: Their Mobility and Relevance to the Study of Laterite Genesis

I. IRON

This element is present in laterite in various forms, the oxide haematite (Fe_2O_3) and the hydrate goethite [$\alpha Fe_2O(OH)$] being the most common. Lepidocrocite has been identified (d'Costa *et al.*, 1966; Webster, 1965); also maghemite (γFe_2O_3) (Bonifas, 1959; Faniran, 1970a), and residual iron oxides such as magnetite (γFe_3O_4) and ilmenite ($FeTiO_2$). Other forms of iron were also suggested in the early literature (Holland, 1903), particularly limonite. Limonite ($Fe_2O_3.H_2O$) was formerly believed to be a crystalline mineral, but was subsequently described as being amorphous (Holmes, 1945). It is now recognized as a mixture of hydrated oxides, predominantly cryptocrystalline goethite aggregates which retain variable amounts of water, and amorphous material. Little is known about this amorphous material. Some may consist of a ferric hydroxide sol sorbed on the basal plane of kaolinite (Follet, 1965; Fripait and Gastuche, 1952) and on gibbsite (Follet, 1965). Alexander and Cady (1962) have suggested that the study of amorphous iron has been neglected and that laterite hardening may be explicable in terms of its behaviour.

In general, the degree of hydration of these oxides depends upon the environment of formation. Thus, limonite usually forms in the lower horizons of a monogenetic profile (e.g. Frasché, 1941; Ollier, 1959) and yellow earths in areas of higher rainfall than red earths (Calton, 1959).

Ferric iron is very stable in the oxidizing conditions of a normal soil (Craig and Loughnan, 1964; Deb, 1949) and is therefore practically immobile (Harden and Bateson, 1963; Pickering, 1962), but it may move to a limited extent as a colloid or organic complex ion (Cornwall, 1958; Pickering, 1962; Harden and Bateson, 1963). Large-scale movement of iron appears to be restricted to the potentially mobile ferrous form (Pickering, 1962). Ferric iron can be mobilized by reversion to ferrous iron under waterlogged conditions (Islah and Elahi, 1954) or reduction by soil bacteria (Bromfield, 1954). Generally ferrous iron goes into solution if the pH is less than 5.5 (Cornwall, 1958; Van Bemmelen, 1941; Britton, 1925), but experimental data provided by Pickering (1962) has suggested that it may be residual even with the pH as low as 4. In neutral or alkaline conditions it is believed to be stable (Grim, 1953). In tropical environments the acid conditions provided at the surface by CO_2-enriched rain-water or by vegetation appear not to be maintained with depth in the profile, and neutral or alkaline soil waters may in fact be the norm (Grim, 1953; A. Mehlich, personal communication). Thus, under the aerobic conditions provided by the soil, iron is likely to be stable except perhaps in the surface horizons. Even in soils which have weathered to the extent that the 2 : 1 clay minerals have broken down and an acid soil environment, rather than that visualized by Grim (1953), has developed, it appears that iron is stable, for the horizons in which kaolin develops are frequently seen to be the horizons in which iron precipitation as mottles and concretions first occurs in the profile (e.g. Craig and Loughnan, 1964; McFarlane, 1969, Ch. 23). In general therefore, the aerating conditions in the soil favour stability or only limited mobility of iron within the pH ranges which occur.

Nevertheless, it appears that iron may be mobilized in the soil over a wider range of pH if it is protected by humus (Cornwall, 1958; Deb, 1949; Schnitzer and Skinner, 1964; Mattson, 1941). Moore and Maynard (1929) demonstrated the effectiveness of iron removal by peat solutions, already indicated by Weiss (1910) and Holmes (1914). It seemed therefore that the abundant vegetation in the tropics might favour iron mobility and this is often alluded to (e.g. Fisher, 1958; Van Bemmelen, 1941). However, it appears that the mobility attributed to iron by this protection may have been overestimated, for Bloomfield (1953) showed that although the aqueous leachates of forest litter readily dissolve hydrated ferric and aluminium complex compounds, sorption of the reaction products takes place concurrently with the solution process and the sorption has the effect of inhibiting further solution of the ferric oxide. Moreover, Grim (1953) has pointed out that due to the high temperatures in the tropics organic matter is readily oxidized so that it does not accumulate and downward seeping waters carry very little organic acid. It seems therefore that protection by humus is unlikely to provide the means by which large-scale movement of iron can take place through the aerated and neutral or alkaline horizons which would otherwise favour its stability.

Similarly, if iron is protected by silica it may be mobile at relatively higher pH than normal (Cornwall, 1958; Beckwith and Reeve, 1964; Reifenberg, 1935, 1938). This may seem to provide a more likely mechanism for the large-scale movement of iron, for the high temperatures of the tropics favour the formation of silicic acid (Cornwall, 1958). However, Deb (1949) has pointed out that the quantity of silica needed is such that the upper horizons would be leached of silica long before appreciable quantities of iron could be mobilized.

In short, although iron may be mobilized under various conditions in humid tropical soils, it appears to be characterized by a limited rather than a sustained mobility. Thus its retention in such soils is generally to be expected.

II. ALUMINIUM

This occurs in hydrates of alumina and in aluminosilicates. It was at first generally believed that the hydrates occurred primarily as a mineral, bauxite, $Al_2O_3.2H_2O$ (Rao, 1928; Crook, 1910; Scrivenor, 1932; Martin and Doyne, 1927), possibly accompanied by a monohydrate mineral diaspore and trihydrate gibbsite (Simpson, 1912). Later it was established that bauxite is not a true mineral, but a mixture essentially of diaspore and gibbsite (Scrivenor, 1932). Gibbsite is usually much in excess of diaspore (Holland, 1903; Sivarajasingham et al., 1962; de Weiss, 1954; Warth and Warth, 1903) and sometimes the only mineral recorded (Hanlon, 1945). In modern literature the free alumina content of laterite is described as occurring predominantly in the form of gibbsite [$\gamma Al(OH)_3$ — Maignien, 1966] with lesser amounts of boehmite [$\gamma AlO(OH)$] and diaspore [$\gamma AlO(OH)$]. Amorphous forms also occur, of which the most common is cliachite (Sivarajasingham et al., 1962; Maignien,

1966). These were formerly incorrectly described as alumogels (Lacroix, 1913). They are now recognized as disordered clusters of cryptocrystalline aggregates (Maignien, 1966). The term bauxite is still used loosely to describe unspecified mixtures of hydrates (e.g. McNeil, 1964).

Aluminosilicates occur in variable quantities in laterite. They occur as kaolinite [$(OH_4)Al_2 Si_2 O_3$ — Maignien, 1966] recorded by Mulcahy (1960), Webster (1965), McFarlane (1969, 1971) and many others. The alumina content of kaolin is 39.5% (Grim, 1953). A more hydrated member of the kaolin family, halloysite also occurs (Watson, 1965; d'Costa et al., 1966).

The hydrates are widely believed to form by desilicification of the silicates. However, there are also some references to gibbsite formation preceding kaolin formation (Harrison and Reid, 1910; Watson, 1965; Young and Stephens, 1965) although it is not always clear if the kaolin is formed by resilicification of the gibbsite or from amorphous material, allophane (Watson, 1965).

Alumina is generally less mobile than iron. Like iron it is most likely to be lost from a soil under acid conditions (Grim, 1953; Prescott and Pendleton, 1952; Hanlon, 1945), acid rain-water being sometimes suggested (Rao, 1928; Van Bemmelen, 1941). However, even more acid conditions are necessary than for the removal of iron (Prescott and Pendleton, 1952; Cornwall, 1958; Hanlon, 1945). Although alumina is normally lost less readily than iron (Grim, 1953; Prescott and Pendleton, 1952; Craig and Loughnan, 1964), residual accumulations of bauxite occur less commonly than those of iron. Wolfenden (1961) has listed the conditions under which it might be expected to accumulate as bauxite. These include profuse rainfall, abundant microflora and free drainage, all of which favour a high pH. Like iron, alumina is provided with added mobility by the protection of humus (Mattson, 1941; Van Bemmelen, 1941; Hanlon, 1945; Schnitzer and Skinner, 1964), although the known removal of ferric iron from some bauxite-forming environments presents some problems (Wolfenden, 1961). Within a given profile, iron and alumina may behave antipathetically, increase of iron content being matched by decrease of alumina and vice versa.

III.   MANGANESE

Manganese occurs in laterite in various mineralogical forms, lithiophorite and birnessite being most common, and hollandite and tadorakite less so (Taylor and McKenzie, 1964). It frequently occurs as black nodules and concretions (d'Costa *et al.*, 1966), usually smaller and less regular than the iron concretions (Cornwall, 1958; Taylor and McKenzie, 1964), and also as veins, coatings, linings and stains (Taylor and McKenzie, 1964; Newbold, 1844; Nagell, 1962; McFarlane, 1969, 1971). In several respects manganese behaves similarly to iron in soils. Both are readily oxidized, form soluble organic complexes (Heintze and Mann, 1947), and may be precipitated by bacteria (Thiel, 1925). However, in other respects they behave differently and manganese concentration does not always directly reflect iron concentration (Andrew, 1954). Sato (1960) has shown that within the normal range of pH in the weathering environment iron is readily precipitated as ferric hydroxide by atmospheric oxidation while manganese ions tend to remain in solution. Some laterites are sufficiently enriched by manganese to be ores (Fermor, 1911; Fisher, 1958; de Chételat, 1947; Nagell, 1962; Chowdhury, in Maignien, 1966), while others have appreciable but non-economic concentrations (Holmes, 1914; McFarlane, 1969, p. 421), or only slight concentrations (de Vletter, 1955). Low manganese contents (Simpson, 1912) and even slight losses of manganese relative to the amount in the parent material (Oertel, 1956) have also been recorded in laterites.

In an untruncated profile, the manganese content often increases towards the surface (Tiller, 1963; de Vletter, 1955; Burridge and Ahn, 1965), possibly due to the generally more oxidizing environment, known to be favourable to manganese stability (Burridge and Ahn, 1965). Thus, manganese contents of sands have been observed to be higher than clays (Tiller, 1963) and relatively higher concentrations are usually associated with colluvial material (Cornwall, 1958). Although a high humus content of the surface material favours loss of manganese (Heintze, 1946) plants may under certain circumstances be responsible for its uptake and accumulation in the soil (Tiller, 1963).

Below the surface layers, if there is abundant moisture and acid conditions, manganese readily goes into solution (Nagell, 1962). Piper (1931) discussed the increase of water-soluble manganese produced by the waterlogging of soil. If drainage is unimpeded there may be considerable loss of manganese so taken into solution, but if a barrier be presented to the movement of such solutions, accumulation results (Nagell, 1962). Manganese concentration has been noted towards the base of some laterite profiles (Newbold, 1844; McFarlane, 1969, Ch. 16 and p. 421) evidently for this reason. Similarly Beater (1940) noted higher manganese concentrations in bog ores than in other refractory deposits. Manganese frequently accumulates in ill-drained depressions. Thus Jensen (1911) noted that

"the accumulations of manganese in these (gilgai) is evidence of absence of sub-drainage, for if any escape existed the manganese would be carried away in the faintly acid soil water which may accumulate there".

In short, two loci of manganese accumulation appear to exist: in the topsoil where conditions favour its stability, and at lower levels where a barrier is presented to the escape of manganese-bearing solutions.

IV. TITANIUM

The oxide (TiO_2) commonly occurs in laterites but is only present in small quantities (Goudie, 1973, p. 30). However, although the absolute content is low and may appear insignificant (Pendleton and Sharasuvana, 1946) it is nevertheless appreciable (Simpson, 1912; Crook, 1909; Harrison and Reid, 1910) and often represents a large accumulation relative to the parent material (Tiller, 1963; Oertel, 1956; Harrison, 1911). For example Mulcahy (1960) recorded 0.49% TiO_2 in laterite and described this quantity as large. Since the content of the parent material was only 0.01% it does in fact represent a very large relative accumulation. Similarly Hartman (1955) described a bauxite containing small quantities of TiO_2 compared with the Al_2O_3 and Fe_2O_3 content and concluded that the enrichment in TiO_2 was twice as great as of these other more obvious accumulations. Thus, despite the large accumulations of iron and alumina, some is lost to the system. If calculations of the quantity of rock consumed to provide the concentrates in the laterite are based on these, they can only provide a minimum figure. TiO_2 has been used as a more satisfactory index mineral for such calculations (Hartman, 1955; Tiller, 1963; Harrison and Reid, 1910). However, Rao (1928) pointed out that although Indian laterites have a high titanium content (e.g. Dey, 1942) there is also considerable removal of titanium and calculations based on its assumed absolute stability should be treated with caution, if it is not also established in what form the titanium occurs.

Titanium often occurs as ilmenite (Maclaren, 1906; Rao, 1928; Hartman, 1955; du Bois and Jeffery, 1955; Hanlon, 1945; Harrison and Reid, 1910), generally agreed to be a stable primary form. This can sometimes be shown to increase upwards in a laterite profile, indicating its residual nature (du Bois and Jeffery, 1955; Hanlon, 1945). However, some laterites contain little or no ilmenite (Simpson, 1912). Other forms include metatitanic acid, $TiO_2 xH_2O$ (Simpson, 1912; Hanlon, 1945), rutile (Rao, 1928; Hartman, 1955), anatase (Rao, 1928), sphene (Rao, 1928), titaniferous magnetite (Rao, 1928), titaniferous haematite (Hartman, 1955) and leucoxene (Hartman, 1955), some of which are secondary forms (Rao, 1928; Craig and Loughnan, 1964). Relatively little is known about the mobility of titanium but some conditions under which it is mobile have been discussed by Craig and Loughnan (1964).

Thus, since it may be lost like the other mobile constituents the absence of accumulations of titanium in laterites need not argue against a residual origin. Conversely, an accumulation of titanium is not evidence of a residual origin unless it is shown to be a primary resistant form, as in the case of Hartman's (1955) study.

V.   ZIRCON

Allied to titanium in its properties (Read, 1947), zircon also accumulates in some laterites and may be regarded as a resistant index mineral (Tiller, 1963). Mulcahy (1960), for example, recorded up to 770 p.p.m. in a laterite, and even the pallid zone showed marked concentrations. The ratio of iron to zirconium showed that there was an absolute loss of iron from all weathering horizons, even from the laterite.

VI.   CHROMIUM

This is an extremely resistant mineral (de Vletter, 1955) and the small quantities recorded in laterite by various authors (in McFarlane, 1969, Appendix 1), often represent appreciable relative concentrations. Iron and chromium have been observed to behave similarly in the profile (Frasché, 1941; McFarlane, 1969, Ch. 23) both rapidly increasing upwards as silica and magnesia are lost from the profile. Frasché (1941) noted a decrease again in the surface horizons. In contrast, Blondel (1954) noted that chrome varies with depth inversely to iron. However, he also observed chromium to decrease down the profile. He suggested that this antipathy of iron and chrome was also expressed in the topographic variations of the laterite, the higher, iron-rich parts of the terrain being apparently poorer in chrome. This agrees with data from Busia in Uganda where, in two profiles studied, chrome was found to be more concentrated in the topographically lower position, the laterite being a residual pedogenetic laterite which has moved mechanically downslope during its accumulation (McFarlane, 1969, Ch. 23). De Vletter (1955) found greater concentrations of chrome towards the bottom of a profile and suggested that the chrome had worked its way down the profile due to its high specific gravity. The general role of chrome may be summarized thus. It is a resistant mineral which accumulates progressively upwards in a residual profile. It may accumulate mechanically in low catenary positions and in a mechanically sorted profile may be found more concentrated at the base.

Little is known of the form in which this element occurs, but Simpson (1912) suggested that it occurs predominantly as chromic hydrate and also a small percentage as chromite.

There appears to be no credible evidence to suggest its mobility (de Vletter, 1955) and so this resistant element, the presence of which Simpson (1912) found in every sample for which it was looked, might fulfill the role which titanium was initially credited with, that is, an index mineral for assessing not only the rock consumption involved in lateritization but also the lateral movement involved in the formation of the crust.

VII.   NICKEL

Nickel also accumulates in laterites (Tiller, 1963; McFarlane, 1969, Ch. 23). The quantities recorded by various authors (in McFarlane, 1969, Appendix 1) are often apparently small, but are not infrequently sufficient for the laterite to be a nickel ore (Fisher, 1958; Mitsuchi, 1954; de Vletter, 1955).

Nickel occurs as hydrated nickel silicate (Fisher, 1958; de Vletter, 1955) and is concentrated near the bottom of the profile (Frasché, 1941; Fisher, 1958; de Vletter, 1955). Nickel at the top of the profile goes into solution. However, the solutions do not travel far, but upon reaching an horizon where magnesia and silica are available, at the base of the profile, the nickel is redeposited (Fisher, 1958; de Vletter, 1955). De Vletter (1955) observed of Cuban laterites that "This interrelation between solution and redeposition is a continuous process; thus, this enriched zone moves gradually downwards". His account of the development of these laterites presents a very strong case for their residual origin, and it is extremely interesting that although the accumulated material, the nickel, is not a resistant mineral, it has nevertheless accumulated as a *residuum*. Such accumulation, as a residuum, of material which has a limited mobility recalls the earlier suggestion that laterite is a residual precipitate (pp. 51–52) which develops with the reduction of a landsurface. The Cuban laterite is in effect a residual precipitate.

VIII. COBALT

Cobalt also accumulates in laterites (de Vletter, 1955; Fisher, 1958; McFarlane, 1969, Ch. 23). Like nickel it readily goes into solution in the topsoil. Its mobility has been studied by Young (1955). Cobalt may, however, concentrate at depth and a similar mechanism to that by which nickel concentrates has been suggested by de Vletter (1955). The horizon of concentration has been observed (Fisher, 1958; de Vletter, 1955) to be slightly higher in the profile than for nickel.

Since, like nickel, cobalt may be lost relatively to iron, chromium and alumina (Frasché, 1941) it cannot be regarded as a resistant index mineral. However, like nickel, it does provide useful information about the mechanism for residual accumulation of mobile elements in the weathering profile.

IX. MOLYBDENUM

Molybdenum is resistant to leaching (Gammon *et al.*, 1954) and is thus another possible index mineral. Since on the one hand its insufficiency in the soil may damage crops while on the other it may accumulate in forage plants to the extent that livestock are poisoned (Gammon *et al.*, 1954), its behaviour in soils has received considerable attention.

It may be adsorbed by the aluminosilicates, halloysite and kaolin (Barshad, 1951; Jones, 1957) and also by hydrous oxides of iron and aluminium (Jones, 1957; Reisenauer *et al.*, 1962). It also occurs in combination with the organic fraction in the soil (Barshad, 1951; Reisenauer *et al.*, 1962). The hydrous oxides of iron are most effective in adsorption (Reisenauer *et al.*, 1962; Jones, 1957). Wells (1956) recorded 10 p.p.m. in ironstone nodules as compared with less than 1 p.p.m. in aluminous nodules. Adsorption varies directly with pH, increasing as the pH decreases (Jones, 1957; Reisenauer *et al.*, 1962; Barshad, 1951). The adsorption also varies directly with the amount of molybdenum and adsorption has the result of raising the pH.

Although molybdenum is known to accumulate in ironstone gravels (Tiller, 1963; Wells, 1956) its role as a resistant index mineral remains largely unexplored.

X.   VANADIUM, COPPER AND OTHER TRACE ELEMENTS

These have been observed to accumulate in laterite. Simpson (1912) and Goudie (1973, p. 29), for example, refer to vanadium accumulations. However, little data is available on the nature of the accumulations.

XI.   PHOSPHATES

In acid soils, these tend to be dissolved out and lost (Cornwall, 1958). However, phosphate may be adsorbed by various forms of active iron (Bromfield, 1965; Martin and Doyne, 1927; Follet, 1965) and aluminum (Martin and Doyne, 1927), the latter being the more effective (Bromfield, 1965). Thus, P_2O_5 content of laterites is often not as low as might be expected from their acidic nature (in McFarlane, 1969; Appendix 1) and it may even accumulate to a certain extent (Mulcahy, 1960; Mulcahy and Hingstone, 1961). Beater (1940) suggested that phosphorous migrates into concretions and noted higher values in bog ores than other refractory deposits.

XII.   SILICA

Silica occurs in laterite in various amounts and in several forms. It occurs as primary quartz (Benza, 1836; Buchanan, 1807; Doyne and Watson, 1933; de Weisse, 1954; Simpson, 1912) particularly where the parent material is acidic. Some Indian samples exceed 20% quartz (Goudie, 1973, p. 29). Quartz sometimes accumulates in the surface horizons, indicating a residual origin for the laterite (de Vletter, 1955; Harrison, 1911; Doyne and Watson, 1933).

Silica also occurs as aluminosilicates, the kaolins, of which some 46.54% is SiO_2 (Grim, 1953). Formation of kaolins from 2 : 1 clay minerals is often accompanied by commencement of iron precipitation and accumulation (Craig and Loughnan, 1964; Nye, 1955; McFarlane, 1969, Ch. 23), and it may be concluded that the breakdown of the 2 : 1 clay minerals liberates the iron which forms the precipitates. The degree of crystallinity of the kaolins appears to be related to the freedom of drainage of the environment in which they form (Craig and Loughnan, 1964). Thus, Muir *et al.* (1957)

noted better crystal development where drainage is better and observed that waterlogging leads to poor crystallinity, and Webster (1965) observed that kaolin crystallinity decreases downslope in a catena. In Uganda, a higher degree of crystallinity is often found in the pisolithic laterites than in the vermiform variety, which forms in a more hydrating environment (McFarlane, 1969, 1971, 1973). Kaolin formation is inhibited if not prevented by the presence of potash, magnesium and calcium (Cornwall, 1958; McFarlane, 1969, Ch. 23).

Silica also occurs as secondary precipitates (Craig and Loughnan, 1964; Harrison and Reid, 1910), in the form of opal or porcellanite (Fisher, 1958; Litchfield and Mabbutt, 1962; Goudie, 1973). In addition, silica occurs sorbed by sesquioxides (Beckwith and Reeve, 1964).

The mobility of silica still presents problems. High temperatures are believed to favour its removal (p. 40). It was also generally believed that its mobility in the tropics was favoured by the lower acidity of tropical soil solutions. Experimental data by Pickering (1962) appears to contradict this, showing that the amount of silica removed during leaching experiments was much greater in slightly acid solutions than in neutral or slightly basic solutions. Subsequently Krauskopf (1967, p. 194) produced experimental data to show that it is no more soluble in near neutral conditions than in acid. Curtis (1970) on the other hand, suggested that although silica solubility may not vary with pH, that of solid aluminosilicates is certainly enhanced by acid conditions. Pickering (1962) suggested that amorphous silica is substantially more soluble than $Al(OH)_3$ and $Fe(OH)_3$ throughout most of the pH range expected in soil or groundwater and that above pH 5 alumina can be expected to accumulate. Silica mobility appears to offer scope for further research.

Summary

Laterites contain a great variety of materials and the mobilities of these are remarkably varied. Some may be regarded as resistant index minerals, while others are mobile under a great variety of conditions. The surprising accumulation of mobile constituents has in some cases been demonstrated nevertheless to be residual. Often it is the minerals present in small quantity which represent the largest relative accumulations, and the study of their behaviour may show that these lesser constituents in the weathering profile are more informative about laterite genesis than the more obvious accumulations which have been more thoroughly investigated.

10

Laterite Genesis

Two very early models of laterite formation proposed *marine* and *volcanic origins.* These enjoyed little popularity and were readily dismissed by Oldham (1893). A third model, involving *termite activity* has received only very limited support. Erhardt (1951) suggested that one group of laterites might be "cuirasses termi- tiques", and that termites were responsible for the "structure vacuolaire generale". Earlier, Nazaroff (1931) had reported the occurrence of ferruginized termitaria. Subsequently Tessier (1959) also recorded "fossil termitieres", but believed that they are not of general significance for explaining ferricrete structures. Goudie (1973, p. 113) considered that termite channels may aerate the profile, encourage the development of vesicular laterites and promote the oxidizing of the ferrous compounds to form indurated ferricrete. Although it is conceivable that termites might make some small contribution to *induration* of laterites, by aiding their aeration, the belief that termites may actually be the cause of laterite *development* is now largely discredited. Nevertheless, the theory still has its adherents (Osmaston, in McFarlane, 1971). It is possible that termites can contribute to laterite genesis by facilitating the reworking of a soil in which precipitates develop, thereby aiding their concentration at the base of the soil, much as a stone line is developed. Nevertheless, a significant relationship between termite activity and laterite genesis remains undemonstrated.

The early concept of laterite as a residuum (pp. 2—3) seems never to have been precisely formulated, but it appears that essentially the accumulations were attributed to the relative immobility of the constituents. Although they were believed to accumulate as a residuum, nevertheless, two phases of limited mobility of the constituents were implied: First, a short-lived phase when the materials to be concentrated were released from their parent material by weathering processes and re-grouped into relatively immobile precipitates; subsequently, a second mobile phase during the re-solution or alteration of the residuum, probably by groundwaters. The model is depicted in Fig. 17a.

The concept of laterite as a precipitate resulted from recognition of the fact that pallid zones typically underlie laterites. This coincided with an increasing understanding of the several ways in which iron and alumina can be mobilized. Thus it came to be

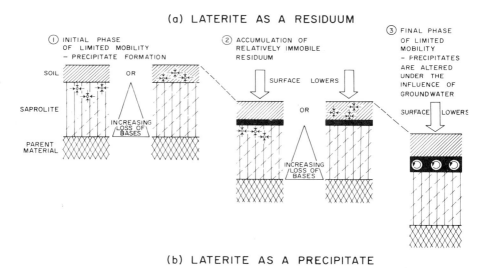

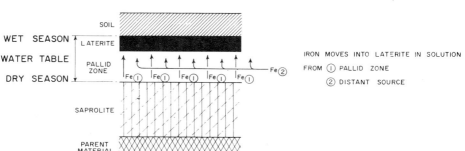

Fig. 17. *Diagrammatic representation of two suggested models of laterite formation.* For further explanation see text.

believed that the enrichment was caused by iron and alumina, in solution, moving into the enriched zones and there being precipitated. The provenance of the concentration was believed to be essentially the underlying pallid zone, but more distant sources were also considered possible if not probable (Fig. 17b).

Two mechanisms were suggested for this enrichment of the laterite horizon, capillarity and the seasonal fluctuations of the water table.

Laterite formation by capillary action was favoured by Maclaren (1906), Harrassowitz (1930), Woolnough (1927) and Holmes (1914). Simpson (1912) described primary laterite as "a true efflorescence, that is, a desposition on the surface of the ground by capillarity, and there deposited as solid matter owing to aeration and evaporation of the water". He implied that it grew progressively upwards, and from this stemmed the general belief that the thickness of a laterite is a measure of the time for which it developed, that is, its age (Holmes, 1914; Maclaren, 1906; Vann, 1963). The evidence against the effectiveness of capillary action in laterite formation is overwhelming (Goudie, 1973, pp. 141—4). Not least is the recognition that capillary action is restricted to a very narrow horizon in the soil (Sivarajasingham et al., 1962; Baver, 1956) seldom exceeding 2 m (6.6 ft). Capillarity, as a major factor, has generally been discredited, but not entirely abandoned as a mechanism by which some duricrusts are believed to accumulate (Loughnan et al., 1962; American Geological Institute, 1962; Stamp, 1961).

The belief that upward movement of iron and alumina was brought about by water table fluctuation was more popular (Pendleton and Sharasuvana, 1946; Pendleton, 1941; Prescott and Pendleton, 1952; Fisher, 1958) and it is still very widely held. Nevertheless there is a growing awareness that this model must be treated with reserve (Sivarajasingham et al., 1962; McFarlane, 1971; Goudie, 1973, p. 145). Essentially the theory is that enriched solutions are carried upwards in the profile with the seasonal rise of the water table and precipitated near the upper limits of the range of fluctuation (Fig. 17b). However, there are a number of seemingly insuperable problems. These are discussed more fully in Ch. 7 but may be summarized thus.

(a) The postulated mechanism cannot explain alumina enrichment since it is inconsistent with principles of chemistry (Sivarajasingham et al., 1962). It also fails in practice to explain upward movement of iron. By the Ghyben—Herzberg hypothesis (in Goudie, 1973) fresh groundwater "floats" on the underlying solutions, thus presenting a barrier to the rise of these solutions in the profile. Sivarajasingham *et al.* (1962) discussed this and pointed out that the development of water supplies in Hawaii is based on this principle.

(b) The scale of many laterite profiles is so large (p. 56) that it becomes necessary to invoke seasonal variations far more extreme than those occurring at present (Walther, 1916). The vast water-table fluctuations required to explain thick laterite profiles do not exist.

(c) The concept of laterite enrichment and pallid zone depletion as synchronous complementary processes is faced with:

(i) the not uncommon occurrence of laterite directly on fresh rock or unleached material;

(ii) the occurrence of laterite over extremely thin leached horizons, completely inadequate to account for the enrichment in the laterite;

(iii) the fact that even the deepest pallid zones are quantitatively inadequate to account for the concentration in the crust (Trendall, 1962).

Some of these occurrences may be explained by a contribution made to the laterite from topographically higher positions (Campbell, 1917; Maclaren, 1906), but there remains the problem of the development of thick laterites with inadequate pallid zones, on interfluves (de Swardt, 1964; McFarlane, 1971). Nor can all of these present interfluves be dismissed as examples of relief inversion (pp. 97—100). Where they are underlain by resistant rocks, such as quartzites, it is unrealistic to suggest that these comprised former lowlands which received a contribution from higher lying rocks (Goudie, 1973, p. 146).

(d) Many laterites occur on surfaces of relatively high relief (p. 32). Some of the relief may be explained by modification of the original laterite surface, but some is original (p. 36). Certainly laterite development is not restricted to completely stable water tables in level or near-level topographies, a prerequisite condition for this model.

Recently, Thomas (1974) has favoured a third possible mechanism for the upward translocation of iron. He cited the hypothesis of Lelong (1966), whereby ionic diffusion may remove rock alteration products from deep within the weathering profile. The hypothesis, however, remains unproven.

Two other kinds of model of laterite formation have recently received more attention; detrital models and those which compromise between the original concept of laterite as a residuum and concepts of it as a precipitate.

Studies of low-level laterites, especially those in a pediment situation, or slope bottom crusts, have been numerous. Considerable detail of their genesis is available, provided largely by pedologists (Maignien, 1966). These laterites are predominantly "absolute accumulations" (a term introduced by d'Hoore in 1954) in that they have received a considerable contribution from topographically higher positions (where an older laterite may or may not occur). The contribution may be mechanical or in solution, or both, as summarized in Fig. 18. It is generally agreed that there is also *in situ* development or primary laterite development in these situations so that the laterite is only partly detrital (McFarlane, 1969; Goudie, 1973) (Fig. 19). The distinction between primary and secondary

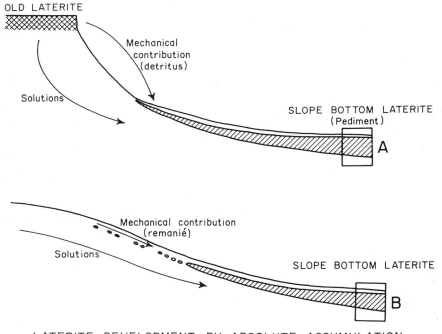

LATERITE DEVELOPMENT BY ABSOLUTE ACCUMULATION

Fig. 18. *Laterite development by absolute accumulation.*
Slope bottom laterites receive lateral contributions from higher topographic positions. These may be mechanical or in solution. Compare the laterite in situations A and B which are essentially similar although A is reputedly "secondary" and B "primary".

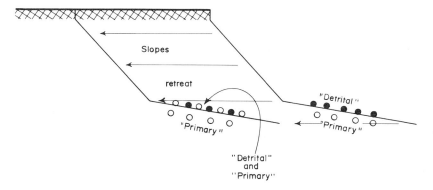

Fig. 19. *Primary laterite development on pediments in association with detrital accumulations.*
With parallel slope retreat, primary laterite developed within the profile of a pediment is incorporated into the surface detrital layers.

laterites is often difficult to make in such situations (Roy Chowdhury *et al.*, 1965), but since detrital elements can often be identified (e.g. McFarlane, 1969) the name "detrital model" may be justifiable. Goudie (1973) has pointed out that this model bears a close relationship to *in situ* models. Pursuing this argument further, since many of the so-called primary laterites are residual and involve the re-solution of material which has mechanically settled or has down-wasted from higher topographic positions, even these are in a sense detrital. It is essentially irrelevant whether the detritus has fallen a few hundred feet down a subcarapace slope onto a pediment or whether it has arrived by a process of mechanical settling and down-wasting through a few hundred feet of relief. Compare situations A and B in Fig. 18. Certainly the distinction between *in situ* and detrital laterite is far from clear.

The development of these so-called detrital or slope-bottom laterites is unique in the discussion of laterite formation in that there is little dispute as to how it occurred. The finer details of the chemistry of the alteration or hydration of such semi-detrital laterites may still present problems but on the whole the outline of their development is based on observation and careful study of actual processes rather than on the circumstantial evidence which provides the basis for theories of high-level laterite genesis. Dispute does arise, however, when this model is stretched to explain all laterite occurrences, especially the fossil "high-level" laterites now occupying interfluve positions. Thus it has been observed that such low-level laterites can upon further erosion or lowering of the surrounding unlateritized areas (or areas where the laterite is unindurated) be left standing above the adjacent country. In effect, the relief beomes inverted. Many examples of relief inversion have been provided in the more recent literature (Maignien, 1966, p. 72; Goudie, 1973, p. 45). Clare's account (Clare, 1960) of the development of a laterite "reef" from a catenary soil association provides one example (Fig. 20). Brown (1968) has described laterites which occur as long sinuous ridges and may have formed as valley laterites. Summit areas where the laterite is unexposed and unindurated can be lowered to leave the formerly lower peripheral areas of exposed and indurated laterite standing relatively higher; for example, the well known soup-plate form of many large remnants of the Buganda Surface (Trendall, 1962) appears to have developed in this way (Fig. 21). Laterite pediments in Uganda can be seen becoming isolated to form lower mesas, as a result of the erosion of the unlateritized surrounding areas (Fig. 22 and Plate 15). Thus it is a demonstrable fact that relief inversion occurs. One may therefore be tempted to explain all the high-level mesas on interfluves in terms of relief inversion, for this is compatible with the one workable model of laterite formation, in the form of the so-called detrital or slope-bottom laterites. This hypothesis must, however, be rejected. There are numerous examples of large areas of laterite blanketing uplands underlain by extremely resistant rocks, for example the quartzites in Kyagwe, Buganda. These areas could not conceivably have formed the original lowlands of the old topography. The outstanding problem remains the explanation of these upland areas of laterite and various attempts have been made to do this by combining the old concept of laterite as a residuum and that of it as a precipitate.

Restating the problem, since these are upland areas and in all probability were the higher landscape elements at the time of laterite formation, a lateral contribution to the laterite must be discredited. Since the underlying pallid zones are inadequate to account for all the concentrates, an overhead source *must* be evoked to explain at least part of the accumulation. To evoke the existence of an older higher-lying laterite, as the source, evades the issue and in practice is unworkable to explain very thick laterite resting *directly* on the highest parts of the landscape. It becomes necessary to assume that the laterite is in part residual, or a "relative accumulation" (d'Hoore, 1954).

This is in effect the compromise which de Swardt (1964) favoured for the development of the groundwater laterites in Uganda. To

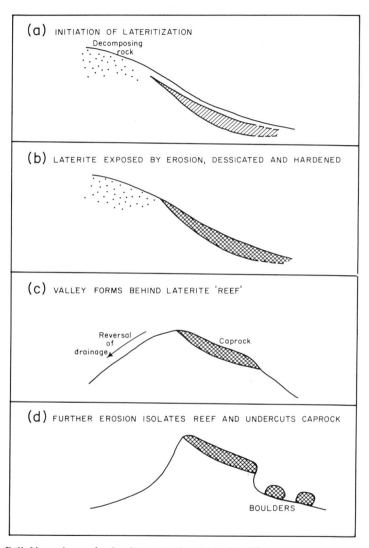

Fig. 20. *Relief inversion — the development of a laterite "reef".*
(After Clare, 1960.)

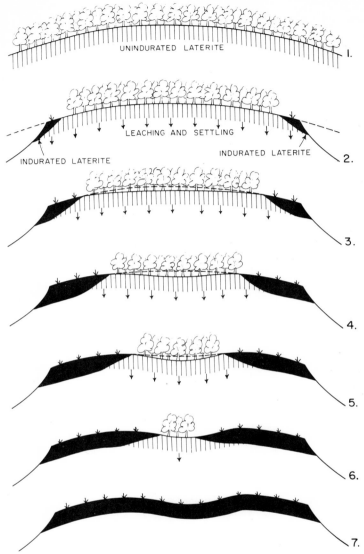

Fig. 21. *Relief inversion — the development of the "soup-plate" form of the Buganda Surface mesas.*
1. A permeable laterite sheet develops under forest cover.
2. Incision and erosion exposes the laterite at the margins of the original interfluve. It becomes indurated and impermeable and the forest vegetation deteriorates. Where permeability is maintained, there is leaching through the carapace and settling of the surface.
3, 4, 5 and 6. Induration of the laterite and vegetation deterioration extend towards the centre of the mesa gradually reducing the area in which settling occurs.
7. Ultimately the entire mesa is deforested and the laterite indurated and impermeable. The central area, subject to the longest period of leaching and settling, now lies relatively lower than the periphery.
(After McFarlane, 1969.)

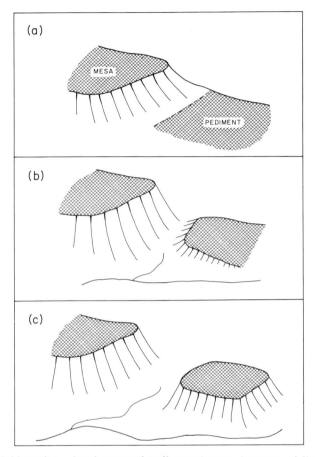

Fig. 22. *Relief inversion — laterite-covered pediments become low mesas following further erosion.*

account for the quantitative inadequacy of the pallid zones in Uganda (Trendall, 1962) he suggested both an overhead source and underlying source and postulated that the profile moved downwards (Fig. 23). He did not, however, move away from the idea that upward enrichment from the pallid zone is brought about by the mechanism of a fluctuating water table. This compromise provides a solution to the problem of the provenance of the concentrates on interfluve areas, reassociating laterite formation with a moving rather than a static profile (p. 65). The laterite develops as the landsurface is reduced. This theory is however still faced with the remaining

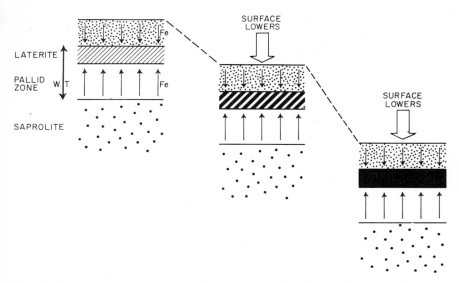

Fig. 23. *Diagrammatic representation of de Swardt's model of laterite formation.*
The laterite accumulation is largely residual, deriving from an overhead source, but an upward contribution from the pallid zone is also postulated. (Following de Swardt 1961.)

problems already listed for the concept of laterite as a precipitate (pp. 57—59), that is, the absence of a plausible mechanism for upward translocation of iron, and the incredible magnitude of the water table fluctuations required. It is clear, however, that the role of the pallid zone is changing. Instead of being a means whereby laterite accumulations can be explained, it is becoming an end in itself, which requires explanation. Thus, de Swardt (1964, p. 317) said

> "an iron-poor pallid zone and a mottled zone generally intervene between the fresh rock and the laterite, and iron *has* to move across these zones in order to augment that in the ironstone above. It thus becomes *necessary* (present writer's italics) to postulate upward as well as downward migration of iron . . .".

Du Bois and Jeffery (1955) suggested that not merely part but all of the enrichment is derived from an overhead source. The laterite is again visualized as a kind of residual precipitate. They suggested that iron goes into solution in the soil and is precipitated on reaching alkaline groundwater (Fig. 24). This theory has the merit that it does not require unlikely large-scale upward movement of iron,

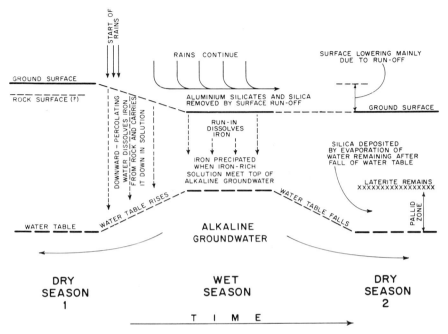

Fig. 24. *Diagrammatic representation of du Bois and Jeffery's model of laterite formation.* The laterite horizon remains stable. Enrichment derives entirely from an overhead source, no contribution being postulated from the underlying material. (After du Bois and Jeffery, 1955.)

leaving the formation of the pallid zone to be explained in terms other than as a complement to laterite formation. Nevertheless outstanding problems remain. Du Bois and Jeffery did not visualize this process occurring in a lowering profile. They believed that the laterite horizon is stable, although the landsurface overhead is reduced chemically and mechanically, and this poses the problem that an unlikely thickness of formerly overlying soil material must be postulated to account for the concentrates which may represent the consumption of up to 500 ft (152 m) of parent material (Trendall, 1962). There is no evidence that laterite formation can begin at such depths. A further objection to this theory concerns the suggestion that iron is mobile in the soil yet stabilized by contact with groundwaters, and du Bois and Jeffery observed in support of this that there is very little iron contained in Ugandan water samples. However, Gear's study (Gear, 1955) of water supplies in Uganda does not confirm the lack of iron in the groundwaters and this is in keeping with the known mobility of iron under anaerobic conditions (p. 82). Du Bois and Jeffery's theory requires iron to be mobile under aerobic conditions and stabilized by contact with the anaerobic zone in which it should be innately mobile.

Trendall's theory of apparent peneplanation (Trendall, 1962) also regards laterite as a sort of residual precipitate (Fig. 25). The main source of enrichment was visualized as being the pallid zone. The entire profile was believed to move downwards as the surface is lowered by wash, and thus the zone of water table fluctuation progressively incorporates new material from which it draws iron for the laterite horizon. This process was visualized as occurring independently on each interfluve. Surface removal of material was believed to expose a peripheral free-face of laterite while the process of laterite formation operates (Fig. 26), and continued lowering of

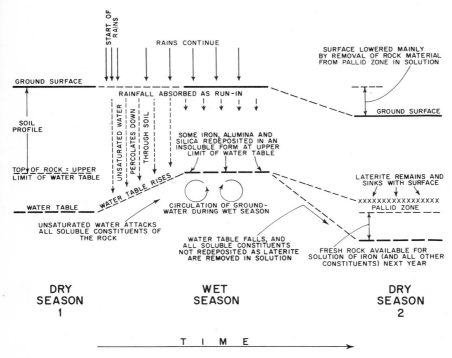

Fig. 25. *Diagrammatic representation of Trendall's model of laterite formation.*
Enrichment is entirely upwards from the underlying pallid zone. Groundsurface, laterite, and pallid zone sink, thereby incorporating progressively more lateritic constituents, so that the total quantity of concentrates in the laterite exceeds the depletion from the existing pallid zone. (After Trendall, 1962.)

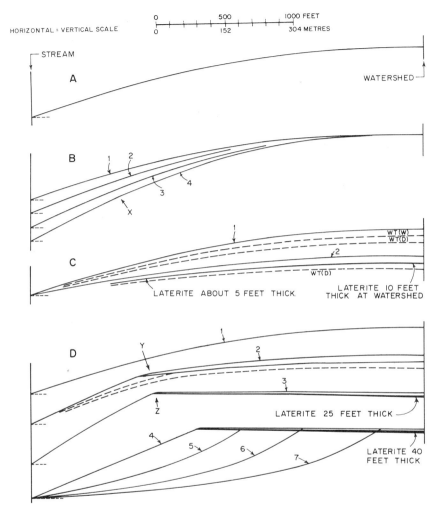

Fig. 26. *Trendall's model of mesa formation in Uganda.*
An interfluve (A) is lowered by surface wash (B). Laterite develops in association with vertical lowering of the surface (C). The combination of these two processes (D) exposes a peripheral free-face of laterite as the laterite develops. (After Trendall, 1962.)

the profile and surface on each interfluve results in the eventual inversion of the relief within this peripheral exposure of laterite. This theory has several merits.

(a) It offers an explanation for the inadequacy of the pallid zones to account for the concentrates in the laterite.

104

(b) It suggests a mechanism for the formation of the inverted summits, or the "soup-plate" form of the laterite mesas of the Buganda Surface.

(c) It offers a reason for the wide variety of altitudes at which laterite mesas occur in Buganda (refer to Plate 16), for the process is a "steadily acting cause which produces a morphological discontinuity" (Penck, 1953, p. 160). He thus considered the Buganda Surface to be an "apparent peneplain", still forming today.

The significance to denudation chronology studies of this last point is enormous. If both laterite formation and the development of the breaks of slope-bounding laterite mesas are acyclic, then the use of such features for denudation chronology analysis and in particular the use of the Buganda Surface as a datum are entirely unfounded.

The theory is, however, faced with the same problems as de Swardt's (1964):

(a) it invokes an impossibly large range of water table fluctuations;

(b) there is no known mechanism for such translocation of iron;

(c) in addition, there is no evidence to suggest that the present-day water table ever reaches as high as the laterite on the mesas.

The merits of this theory have been diminished by more recent studies which suggest:

(a) that the "soup-plate" form of the mesas is a postincision modification of the surface (pp. 97, 99, this book) and

(b) that the apparently haphazard altitude range of the mesas can be resolved into two groups of *in situ* laterite-capped mesas with intermediate detrital laterite-capped mesas (p. 79, this book), the products of modification processes.

The main contribution of Trendall's theory, therefore, remains, like de Swardt's (1964) that it offers an explanation for the quantitative inadequacy of the pallid zone to account for the concentrates in the laterite, again associating laterite formation with a lowering landsurface and lowering profile. For this a strong case exists (Ch. 7).

A further theory of compromise, outlined in Fig. 27, has been evolved in Uganda (McFarlane, 1971). It also regards laterite as a residual precipitate. The groundwater laterite is believed to accumulate as a mechanical residuum during the late stages of reduction of a down-wasting landsurface. The original precipitates form within the relatively narrow range of fluctuation of the groundwater table, which sinks as the landsurface is reduced by erosion. They become

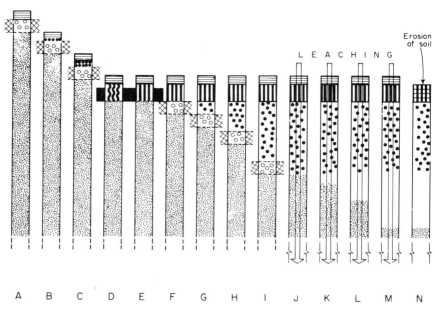

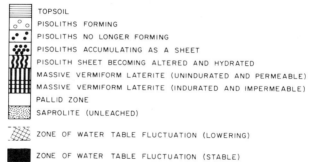

Fig. 27. *Diagrammtic representation of the author's model of laterite formation.*

A. Iron is segregated into pisoliths, within the narrow range of oscillation of the groundwater table, during a late stage of landsurface reduction.

B, C. Zone of precipitate formation lowers as landsurface is reduced and the precipitates accumulate as a sheet at the base of the soil.

D, E. Landsurface reduction ceases and the accumulated sheet of pisoliths becomes altered and hydrated to form a massive sheet of planation surface laterite.

F. Water-table once again begins to lower upon initiation of the succeeding cycle and pisolith formation resumes.

G, H, I. With continued water table lowering, the zone of pisolith formation lowers leaving above it a spread of pisoliths.

J. The water table is ultimately lowered beyond the depth at which pisoliths can form.

K, L, M. Leaching through the carapace depletes the saprolite underlying the spread of pisoliths, forming a pallid zone.

N. Deforestation leads to induration of the laterite and its loss of permeability. Pallid zone formation ceases.

106

incorporated into the lower parts of the soil mantle where they accumulate as an increasingly thick layer (see Plates 10 and 14). When down-wasting has ceased and the water table has stabilized the residuum is hydrated and altered and a massive variety of laterite develops which has every appearance of a true precipitate (see Plate 8). This alteration of a residuum or detrital laterite is very similar to the so-called detrital model discussed earlier (pp. 95—96), but the vertical lowering of the low-relief landsurface allows the detritus to cover thickly all parts of it. Like du Bois and Jeffery's theory (1955) no contribution is believed to have been provided by the pallid zones which underlie the high-level laterites. This theory is in part based on a study of the low-level laterites which can be seen to be at various stages of this sequence of landsurface reduction and laterite development (McFarlane, 1969). Since the high-level laterites are essentially similar, a similar process of development was proposed for them also, and no unlikely magnitudes of water table oscillation are required. This zone is merely a narrow zone in which iron is segregated from the saprolite to form discrete bodies or pisoliths. Large vertical spreads of such precipitates, for example, those found below the planation surface laterite on mesas (Figs 13 and 27 and also Plate 1), are believed not to indicate large fluctuations of water table, but merely the path through which the narrowly oscillating water table migrated as it lowered in response to landsurface erosion. It associates laterite formation with a surface of some original relief and not exclusively to planation surfaces, although the end-product, the mature laterite, is exclusive to very low relief. This model is in fact merely a mechanical variation of Nagell's model for the development of the laterites of Cuba in which the laterite profile lowers and accumulates its residual content by a continuous process of solution in the upper horizons and deposition in the lower (pp. 51—52). Since pallid zones occur below some laterites a causal relationship must exist, and it is suggested that the permeable nature of *in situ* laterites allows the substrata, the saprolite, to be leached after the laterite is incized (pp. 60—61).

The so-called typical laterite profile consisting of massive laterite underlain by spreads of pisoliths or mottles, underlain by pallid zones is thus believed to be the product of two cycles of erosion, (the end of the first and the beginning of the second) and the pallid zone may in fact be currently forming, i.e. acyclic, if the permeability of the laterite has been maintained since the incision.

The great divergence of opinion on laterite genesis which occurs in even the most recent accounts clearly demonstrates the scope for further research in this field. Hopefully it will establish a more satisfactory relationship of laterite genesis to landsurface development than those previously suggested, for the results of the application of the assumed relationship to denudation chronology studies have been far from satisfactory. Studies in Uganda provide a notorious example.

11

Laterite and the Denudation Chronology of Uganda

Laterite has had a significant place in the development of theories of landscape evolution in Uganda. These theories provide a case study of the way in which laterite can be used and misused in the study of the geomorphology of an area.

King (1962) established a pattern of morphological development in South and Central Africa, which he believed to be applicable to all parts of the African continent, including Uganda. He recognized an original planation surface, the Gondwana Surface, the erosion of which he assigned largely to the Jurassic. Arid or steppe conditions were believed to have prevailed during its development. So complete was this planation that it obliterated the post-Karoo folds and faults largely of Liassic age (King, 1962, p. 250). King believed that this cycle was terminated by the rupture of Gondwanaland in the Early Cretaceous, but that even in the Jurassic an arm of the sea had developed along what is now the East African coast. The margins of Gondwanaland subsequently sagged and were buried beneath Cretaceous sediments. Inland, a new cycle of erosion created the post-Gondwana Surface, trimming the older surface. At the end of the Cretaceous further earth movements attributable to isostasy caused the initiation of the African cycle of planation. These earth movements caused marginal uplifting of Africa, so that the Gondwana and post-Gondwana surfaces sloped backwards into the continental interior.

The African cycle of erosion continued through the Tertiary to the Late Oligocene or Early Miocene. The planation surface produced by this cycle is variously described as the Early Cainozoic Surface, the Mid-Tertiary Surface, the Miocene Peneplain and the Sub-Miocene level. After the Early Miocene the development of this surface was interrupted by renewed earth movements. End-Tertiary followed by Pleistocene surfaces developed at its expense.

Remnants of the Gondwana Surface are believed now to be restricted to small residuals on the highest interfluves. The African cycle of planation is considered to be responsible for the many incredibly flat elements in the African landscape. As well as this diagnostic flatness, the African Surface is further characterized by inselbergs (King, 1962, p. 265; Holmes, 1945, p. 192; Handley, 1954; Willis, 1936, pp. 31, 205). Since inselberg development is popularly attributed in Uganda to the exhumation of the irregularities of the basal surface of weathering below a deeply weathered profile and since deep weathering is attributed to a humid rather than an arid or steppe environment, it seems unlikely that such inselbergs can be regarded as residuals of the Gondwana Surface, or rather its basal surface of weathering; such deep weathering may, however, have occurred during the post-Gondwana cycle, for in Zaire this surface is reputed to carry a laterite (King, 1962, p. 257), which would be indicative of significant quantities of available moisture.

In Uganda, Wayland (1921) initially recognized a single peneplain which he called the Buganda Plateau or Buganda Peneplain. Subsequently he recognized two others (Wayland, 1931, 1933, 1934b), an older surface represented on the resistant interfluves in south-west Uganda (Combe, 1932) and a younger surface extensively developed in the lowlands of Kyoga (Location Map). These he called PI, PII and PIII (Fig. 28). This tripartite division appeared to accord roughly with Dixey's findings, supported by King (1962); that is, the three main surfaces of Gondwana, African and End-Tertiary cycles.

It is not possible, however, to equate these three main surfaces in Uganda with those in south and central Africa by morphological continuity. Since the Miocene surface (Wayland's PII) is reputedly the most readily recognized of the surfaces (Willis, 1933, 1936; Pulfrey, 1960) attempts were made to relate this surface to its morphological equivalent elsewhere. Eastwards from Uganda a Miocene peneplain has been recognized in Kenya (Willis, 1936). It has been mapped (Pulfrey, 1960; Saggerson and Baker, 1965) and dated from fossil evidence (Willis, 1936; Pulfrey, 1960; Saggerson and Baker, 1965) although Bishop (1966) has questioned the validity of the dating. However, to attempt to extrapolate from this planation surface to Uganda across a major continental divide

LOCATION MAP

running north-south through eastern Uganda and western Kenya is in any case unwarranted as morphological continuity cannot be assumed (Bishop, 1966). Moreover, the mesas of the Buganda Surface disappear a short distance to the east of Jinja (see Location Map) and although Ollier (1959) believed it to be represented by quartzite summits in eastern Uganda, this surface is not represented further towards the watershed. Thus the Buganda Surface is not morphologically continuous eastwards with a dateable surface in Kenya.

111

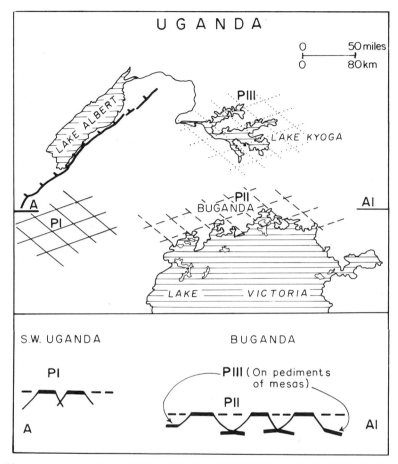

Fig. 28. *Diagrammatic representation of the spacial distribution and altitudinal relationships of Wayland's three erosion surfaces in Uganda.*
(Following Wayland, 1931, 1933, and 1934b.)

To the south, Willis (1936, p. 141) claimed that the Miocene peneplain was traceable from the coast

"into the Iringa Highlands, across the great Ruhaha Valley to the Tanganyika plateau, and thence to the southern shore of Lake Victoria, where it coincides with Wayland's original Buganda plain. This correlation thus rests upon the essential continuity of the planed surface and upon fossil evidence at both the coastal and interior extremes".

Since Lake Victoria is some 150 miles (240 km) wide, this claimed coincidence of the surfaces appears to be based on a rather liberal interpretation of morphological continuity.

Wayland's PIII is extensively developed to the north of the Buganda Surface, in the Kyoga Lowlands, again isolating it from possible equivalents in that direction.

It is from a westerly direction that the most promising attempts have been made to trace the morphological continuity of the Buganda Surface. Lepersonne (1956) described a multiplicity of surfaces in Zaire which he attributed to three major peneplains. He described a Cretaceous surface at 1,700—1,900 m with vestiges of a higher surface (?Gondwanaland) at 2,000—2,200 m. King (1962, p. 257) said that these are "matched" in western Uganda by morphologically similar surfaces at about 1,420 and 1,650 m respectively. However, the surfaces are not continuous, as the Western Rift Valley intervenes (Fig. 29). Ruhé (1954, 1956) has shown that Lepersonne underestimated the effect of warping and faulting towards the rift shoulders, and that the three peneplains recognized by the latter on the Zaire side of the Rift are one and the same warped and faulted

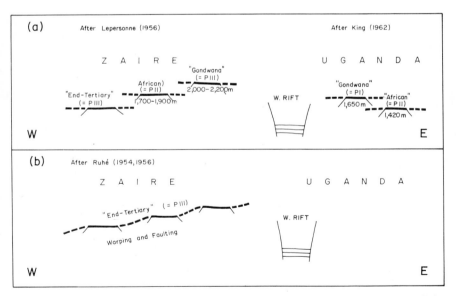

Fig. 29. *Interpretation of the relationships between laterite-capped mesas in eastern Zaire and western Uganda.*
(a) Following Lepersonne (1956) and King (1962).
(b) Following Ruhé (1954, 1956).

End-Tertiary surface, or PIII (Fig. 29). Although Lepersonne's account was approved by Dixey (1956), Ruhé's belief that it is PIII and not PII which is represented on the shoulders of the rift has been supported by Hepworth (1961), de Swardt (1964), de Heinzelin (1963), Gautier (1965, 1967), Laruelle (1961) and Bishop (1966) although Ruhé's dating of it is certainly incorrect (Dixey, 1956; Bishop, 1966). Once again the Buganda Surface (PII) is isolated from possible equivalents.

In short, the identification and dating of the three main surfaces recognized by Wayland in Uganda cannot be confirmed by tracing the morphological continuity of even the most readily recognizable surface, the Buganda Surface (PII), with equivalents identified and dated elsewhere by sedimentary evidence.

Within Uganda there is no direct stratigraphical evidence of the age of the Buganda Surface. This is largely an area of erosion, and dateable sediments are limited. They occur only in deeply down-faulted troughs and also below the volcanics on the eastern boundary.

Karoo sediments are now known to occur in five localities (R. Macdonald, personal communication). Best known and apparently typical of them (Bishop, 1966) is the graben at Entebbe, bounded by faults approximately one mile (1.6 km) apart and containing Karoo sediments certainly 1,000 ft (305 m) and possibly 1,400 ft (426 m) thick. These sediments provide the only means of determining a lower age limit for the Buganda Surface, which is planed across them. However, as noted earlier, the Gondwana Surface also obliterated such Karoo faults.

No sediments have been found indisputably overlying the Buganda Surface, but in two localities Miocene sediments occur on surfaces which were believed to be down-faulted parts of the Buganda Surface.

Thus, within the Kavirondo graben, lower Miocene fossiliferous sediments were found resting on "an old laterite ironstone and filling slight depressions in the old surface" (Shackleton, 1951). These were correlated with gravels under the Gwasi (Kisinguri or Rangwa) volcanic deposits, and assigned to the Sub-Miocene peneplain (Shackleton, 1951, p. 377). Although the laterite-capped Buganda Surface mesas occur some 100 miles (160 km) north-west of Kisumu, the term Buganda peneplain was used by Shackleton in Kavirondo and the sediments were here taken to indicate that the Buganda Surface was rightly supposed to be the Mid-Tertiary African Surface. Bishop (1966) has discussed the fallacy of this argument and recalled that King (1957) observed that there is in fact a difference of opinion as to whether the main platform of the region passes under the sediments or truncates them (Dixey, 1945; Kent, 1944; Shackleton, 1951). Bishop further pointed out that these lower Miocene sediments, overlain by volcanics, rest on a very irregular topography rather than a subdued and recognizable landsurface. The equation of the Sub-Miocene surface of Kavirondo with the Buganda Surface is now largely discredited (de Swardt and Trendall, 1970). Certainly, apart from the fact that both bear a laterite of some sort, there seems no evidence to support their chronological equation.

In the Western (Albertine) Rift Valley, sediments now acknowledged as broadly Miocene (Hooijer, 1963; Leakey, 1967) were found to overlie an erosion surface bearing a laterite (de Heinzelin, 1963; Bishop, 1967). Again the equation of this surface with the Buganda Surface appears to be largely unsupported (de Heinzelin, 1963; Gautier, 1965, 1967), and as in Kavirondo, the occurrence of a laterite appears to have been considered significant for identifying the surface (de Swardt and Trendall, 1970). Laterite also occurs on PIII, but this seems to have been ignored. An examination of the Sub-Miocene laterite in the Western Rift by the present writer shows it to be an immature laterite much more akin to those occurring on PIII than to those on the mesas in Buganda.

It seems therefore that in the absence of either dateable sediments resting on the Buganda Surface or of the possibility of tracing the morphological continuity of the Buganda Surface with other dated surfaces elsewhere, the identity of the Buganda Surface rests largely on the general grounds of its relationship with other surfaces of the surrounding area within Uganda.

It is not surprising that Wayland (1921, 1933, 1934b) at first recognized only a single peneplain in Uganda. The Buganda Surface, particularly in the Kampala area where the residuals are small, gives a very strong impression of extreme flatness (see Plate 16). Since Wayland (1931) subscribed to the belief that laterite is a precipitate which forms after the development of a peneplain, the fact that the Buganda Surface carried a thick massive laterite meant that it *must* have "almost no relief". The reputation already acquired by the African Surface for being the most obvious morphological feature in the African scene may have been the main reason for Wayland's belief that the Buganda Surface was to be equated with the African Surface. The later recognition of remnants of what was believed to be another surface preserved at higher altitudes than the Buganda Surface over the resistant rocks on interfluves in Ruanda and Ankole

115

in south-west Uganda, together with the recognition that the pediments of the Buganda Surface were watershed representatives of a younger surface more extensively developed further north in the Kyoga Basin, placed the Buganda Surface centrally in a tripartite sequence. This established the Buganda Surface more firmly in the "African niche", for its relationship with these other surfaces of the area appeared identical with the relationship of the African Surface to the Gondwana and End-Tertiary surfaces elsewhere.

Furthermore, although the significance of absolute altitude is generally denied (King, 1962, p. 235), the fact that in many parts of south and central Africa the African Surface occurs at about 4,000 ft (1,219 m) (King, 1962, pp. 223, 235) and the Buganda Surface occurs at 4,300—4,400 ft (1,311—1,341 m) (Pallister, 1960) is a coincidence which must have been difficult to ignore entirely. In total there seemed convincing circumstantial evidence for suggesting the equation of the Buganda and African Surfaces.

In detail, PI was believed by Wayland (1934a) to be the only true peneplain present in Uganda (Fig. 30). It was thought to differ from the other two in that there are only patches of laterite on it (Wayland, 1934a) and that it has an undulating and uneven surface unlike the "amazingly flat" laterite surfaces. It was reputedly underlain by considerable depths of saprolite, and upon rejuvenation the succeeding surface (PII) cut back from the rivers into this saprolite to produce the "amazingly flat" surface on which the laterite developed. This process of surface development was called *etching* and PIII was believed to have formed in a similar manner (Wayland, 1934a). Since Wayland believed the Indian distinction between high-level *in situ* and low-level detrital laterites to be applicable to Uganda (Wayland, 1935), the *in situ* laterite of PII was regarded as diagnostic of it. An *in situ* laterite came to be a major criterion of recognition of the African Surface (Ruhé, 1956; Johnson and Williams, 1961; de Heinzelin, 1962; de Swardt, 1964, p. 322). Perhaps because the distinction between *in situ* and detrital laterite is far from clear (pp. 35—36) the diagnostic criterion lapsed in practice to "laterite" rather than "*in situ* laterite". This general equation of well-developed duricrusts with Tertiary surfaces is widespread (Goudie, 1973, p. 90). Goudie has made the valid point that this may be simply because the older crusts have not been recognized as such. Nevertheless, on the whole, it is a popular practice to assign well-developed crusts to the Tertiary (Vann, 1963), in agreement with Wayland's conclusions.

In Uganda and elsewhere in East Africa the occurrence of laterite

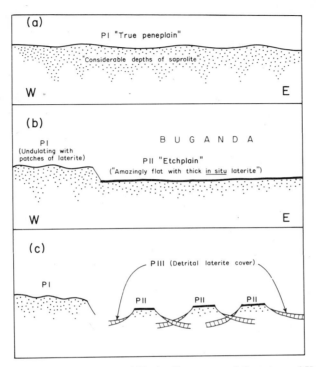

Fig. 30. *Diagrammatic representation of Wayland's concept of the nature of Uganda's three erosion surfaces.*
(Following Wayland, 1934a.)

assumed disproportionate significance. Thus, for example, Stockley and Williams (1938), noting that the laterite on the Miocene surface is diagnostic of it, described this surface in Karagwe as an

"old age of erosion consisting of an alternating series of wide open valleys separated by ridges rising generally to a height of 1,000 ft above the valleys and a drainage pattern perfectly adjusted to structure".

It seems that the laterite is diagnostic of a surface which by no criteria qualified for the name planation surface (p. viii)! Very far from "amazingly flat" it is doubtful if this surface would ever have been assigned to the Miocene in the absence of a laterite cover. Similarly, the fact that laterite underlay the Miocene sediments in the Western Rift Valley and in Kavirondo had strong implications as to the

117

identity of the surfaces. In the case of the latter, the laterite is a "basal gravel of quartz and ironstone" (Willis, 1936, p. 149) and in the former a "complex assemblage of concretions and gravels" (de Heinzelin, 1963), hardly massive *in situ* laterites, but this seems to have been overlooked in favour of the fact that they are laterites.

In detail it can be seen that the nature of these three surfaces is not in all respects similar to that of the assumed counterparts elsewhere in Africa. For example, as well as the characteristic flatness, the African Surface usually has inselbergs rising above it. Yet it is PIII which is characterized by inselbergs in Uganda. Ollier (1959, 1960) has described their development there, by the exhumation of the irregularities of the basal surface of weathering of the Buganda Surface profile. Furthermore, the Gondwana Surface, which is equated with PI, reputedly developed under arid or steppe conditions which do not favour deep weathering. Yet in Uganda it is described as deeply weathered, so that PII was etched out of this surface. Also in view of this characteristic deep weathering, its preservation on resistant interfluves (Willis, 1936, p. 36) is not easily understood unless the deep weathering was not ubiquitous. If this is the case, then the extensive development of PII over the quartzites of Kyagwe, further towards the continental interior is difficult to understand.

Apart from these anomalous aspects of the reputedly diagnostic nature of the surfaces, difficulties arise with their relationships. In the type area of PII (Kampala), the relationship of PII to PIII was broadly established. The pediments of the Buganda Surface (PII) were generally to be equated with Wayland's PIII. However, the type area of PII is far removed from that of PI in the west, and the intervening area presented problems. The typical laterite-capped mesas of the Buganda Surface are not continuously represented westwards. Areas in which mesas occur are separated by lowlands over the weaker lithologies. The groups of mesas thus separated are in fact found to occur at different altitudes, and although their form is apparently similar to that of the Buganda Surface mesas in the type area, so well described by Pallister (1951, 1953, 1954, 1955, 1956a,b,c,d, 1957, 1959, 1960) there arose the question of whether these were the same surface or a flight of similar surfaces. Thus, McConnell (1955) maintained that the laterite-capped mesas in the Koki area, occurring at 4,700–4,800 ft (1,433–1,463 m) represent an older surface than the Buganda Surface, with mesas at 4,300–4,400 ft (1,311–1,341 m) (Fig. 31a). He based his conclusion on Hatton's mapping in Masaka (Hatton, 1953) which showed the presence of a scarp separating the Buganda Surface from the Koki Surface. Pallister (1956b,d) believed them to be the same surface up-warped to the west. Warping had been suggested to explain local differences in the altitudes of mesas (Johnson, 1954, 1956, 1960; Pallister, 1960) and such a mechanism seemed reasonable to explain interregional differences. Pallister's conclusion was based on his study of the stages of destruction of the surface. He had observed that while the laterite is intact there is parallel retreat of the subcarapace slopes which often maintain a remarkably constant angle of 24–27° (Pallister, 1957). On destruction of the carapace there is progressive down-wasting so that the summit is lowered and the angle of slope reduced. Thus, from a study of the down-wasted residuals between Koki and Buganda, Pallister (1956b,d) concluded that the Buganda and Koki surfaces were to be equated (Fig. 31b). Johnson (1959) also believed these surfaces to be the same, but disagreed with Pallister's suggestion that the difference in altitude is due to upwarp. He believed the differences to be original and due to varied development of the surface over lithologies of different resistances (Fig. 31c). This was supported by Doornkamp and Temple (1966).

However, even *within a limited area* two or more laterite levels have been observed. Thus, in the Singo area over the Singo Series relatively flat-lying laterites occur at about 5,000 ft (1,555 m) (Johnson and Williams, 1961), some 300–400 ft (91–122 m) above the laterites of the Buganda Surface developed on the Mityana Series (Buganda Series) surrounding the Singo Highlands (Fig. 32d). Also, Plummer (1960) observed benches of laterite flanking laterite-capped mesas in the Rwampara Mountains (Fig. 32a), and in Kampala, the type area of PII, Harris (1947) pointed out that although individual summits may be very flat, the summits do not fall within a narrow altitude range (Fig. 32h), and suggested that three intermediate surfaces occur as well as summits attributable to PII (Fig. 32i). Wayland (1934a) himself drew attention to the fact that PII appeared to be any of three surfaces locally expressed at the expense of the other two (Fig. 32j).

Thus, it appeared that the Buganda Surface might not in fact be a single surface. Some still maintained that it was and various reasons were given to explain the evidence to the contrary. For example, Pallister (1951, 1954) disputed the validity of Harris' (1947) conclusions on the grounds that in the area in which he studied the Buganda Surface, that is, the type area, the mesas are too few and far between to determine whether or not they belong to one or several surfaces. Plummer (1960) did not believe that the benches of laterite

119

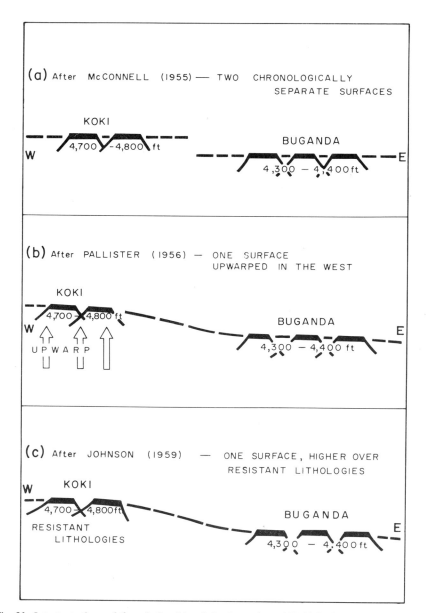

Fig. 31. *Interpretations of the relationship of the Buganda and Koki Surfaces.*
(a) Following McConnell (1955); (b) following Pallister (1956d); (c) following Johnson (1959).

he had observed indicated two chronologically separate laterite developments. He said that since the laterite on the benches and on the mesas could in one instance be seen to be linked in a continuous sheet, this indicated their common age (Fig. 32b). He argued that since a sharp break of slope is not conducive to lateritization, the discontinuity between benches and summit flats was understandable. Doornkamp and Temple (1966) also subscribed to the belief that only one peneplain exists in the Rwampara Mountains "for in places the two supposed levels can be seen to merge imperceptibly one into the other" (Fig. 32c). However, the argument that a continuous laterite sheet indicates a chronological entity has been shown by more recent studies in Buganda (McFarlane, 1969, 1971) to be entirely incorrect. In Kyagwe, laterite plotted by Hepworth (1951, 1952a) occurs at all levels from about 4,400 ft (1,341 m) to below the level of Lake Victoria at 3,780 ft (1,152 m) (Fig. 32m). Nevertheless, the long, continuous sheets of laterite, which appeared in this area to indicate that the mesas belong to one high relief surface (upland surface) have been shown to be diachronous. The higher mesas are formed of two chronologically separate *in situ* massive vermiform laterites, frequently linked by detrital pisolithic laterite sheets (Fig. 32n and Plate 17). Each of the two surfaces identified by their distinctive vermiform laterite cover represents a period of stability unsurpassed by succeeding cycles and there is no case for "lumping" them into one single peneplain or "upland surface" simply because they are linked by detrital laterites where the intervening erosional slopes are shallow. It seems likely therefore that more than one surface is preserved also on the Rwampara Mountains, but only a study of the nature of the laterite can satisfactorily demonstrate this.

To explain the situation at Singo, it has been suggested that laterites formed synchronously at different levels (Fig. 32e), due to the different resistances of the lithologies (Johnson, 1959; Johnson and Williams, 1961). This is in effect a resurrection of the early belief that laterite may form synchronously on plateaux and lowlands (p. 31). Such an explanation raises the further question of why there was not a comparable development of laterite on PI (p. 116). Furthermore, a study of the nature of the high-level laterite in Kyagwe (McFarlane, 1973) has shown that although a thin veneer of pedogenetic laterite may still be forming on the surfaces of the mesas as the vegetation deteriorates (and this should not be confused with the induration of the underlying groundwater laterite which accompanies it), the main body of laterite is a groundwater variety, which in its present position, perched far above the groundwater table, cannot possibly still be forming (see Plate 4). The alterations it is undergoing can only be described as destructive (McFarlane, 1969). As further evidence against the suggested synchronous development of the laterites at different levels at Singo, P. Brock (personal communication) has pointed out that the streams of the Kyoga-Kafue drainage system have graded profiles extending up into the Singo area and that if these recent streams can attain grade it seems unlikely that the streams active during the Buganda cycle should have left this highland so undissected that flat-lying laterites could develop significantly above the general level of the Buganda Surface. Two chronologically separate laterite developments are thus suggested (Fig. 32f).

Another explanation for the differences in altitude of the laterites has been offered by Bishop and Trendall (1967) who suggested that the laterite may have been at the same level originally, but that differential postincision settling over different lithologies might have caused the laterite levels to drift apart (Fig. 32d). However, although postincision settling and the development of pseudo-karst topographies on laterite have been described from many areas (see Goudie, 1973) the result is consistently to increase the local relative relief rather than uniformly to lower large areas which maintain their near-level aspect. Postincision settling has also been studied in Buganda (McFarlane, 1969) and the evidence does not support widespread regional lowering of the laterite surface but rather the development of minor surface irregularities; enclosed hollows and marginal cambering of the carapace being typical.

The more detailed observations of the Buganda Surface, which showed that some laterite sheets are continuous through an altitude range in the order of 500 ft (152 m), led to the suggestion that it may be a single surface of high relief (Hepworth, 1951, 1952a,b). De Swardt and Trendall (1970) suggested that the original relief may have been in the order of 900 ft (274 m) (Fig. 32k) and this was supported by Doornkamp and Temple (1966). In Hepworth's area, Kyagwe, attempts were made to reconstruct this proposed single high relief surface (McFarlane, 1971) and the result was an improbable stepped laterite-covered landsurface. Taking this result in conjunction with the evidence provided by the nature of the laterite, it was concluded that this apparently single high relief surface represented two modified planation surfaces, obscured by linking sheets of detrital laterite (Fig. 32n). This explanation for the apparently high relief of a single laterite surface may apply further to the west and it certainly avoids the problem posed by the interpretation offered by de Swardt and by Doornkamp and Temple; that is the problem of how a laterite can develop on a surface of such high relief, for although a certain relief is in accordance with laterite developing, with the landsurface, as a residuum, the relief they cite is certainly too great to be accommodated by such an explanation.

Trendall's attempt (Trendall, 1962; see also pp. 103—105) to explain the development of the Buganda Surface mesas in terms of the acyclic development of laterite on individual mesas was a further attempt to explain the puzzling problem of why laterite, widely presumed to be a planation surface development, should occur at such a variety of altitudes (Fig. 32k) on what is reputedly the world's best example of a laterite-capped planation surface.

Thus the identity of the Buganda Surface has been very seriously questioned, and many of the questions relate to basic problems concerning the relationship of laterite genesis to relief. Is laterite formation restricted to planation surfaces, or surfaces of very moderate relief? If this is so then it is likely that the Buganda Surface is not a single surface but two or possibly three surfaces. Can laterite develop on a surface of some 900 ft (274 m) relief, introducing the possibility that it is a single surface? Can laterite form synchronously at different levels, possibly blanketing surfaces at different ages? Can a single near-level laterite surface be differentially lowered, either during formation or as a postincision modification, to give the impression of two chronologically separate sheets or surfaces? Can laterite form independently on separate interfluves to form an acyclic "apparent peneplain"? Clearly a better understanding of the way laterite forms would assist in answering these questions. Before laterite can be used successfully as a diagnostic criterion in denudation chronology analysis the relationship between laterite development and topographic development needs clarification.

Not only has the identity of the Buganda Surface been questioned, but recent evidence has challenged the age to which it has been assigned on the basis of what were formerly believed to be its easily recognizable characteristics. Ruhé (1954) concluded that the ancestors of the present day Nkusi/Kafu, Katonga and Kagera (Location Map and Fig. 33) rivers drained westwards on the End-Tertiary surface (Wayland's PIII), across the present location of the Albert-Semliki-Edward Rift Valley to the Ancestral Congo. This general pattern of drainage on PIII is widely supported (Cooke, 1957; Radwansky and Ollier, 1959; de Heinzelin, 1962, 1963; Gautier, 1965, 1967). Rifting was at first believed to be of Pliocene

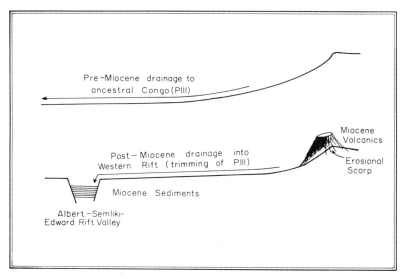

Fig. 33. *Drainage pattern, rifting and surface development in central Uganda.*

age (Ruhé, 1954), but detailed investigations of the rift sediments (de Heinzelin, 1963; Gautier, 1965, 1966, 1967; Gautier and Geets, 1966; Hooijer, 1963; Bishop, 1966, 1967, 1968) have assigned to it a Pre-Miocene age. This means that the so-called PIII, believed to be Late Tertiary, had developed before the Pre-Miocene or Lower Miocene rifting introduced a local base level for the subsequent trimming of this surface in Pliocene and Pleistocene times (Macdonald, 1961; Hepworth, 1961). PIII must therefore be broadly equated with the African Surface and it becomes necessary to move the Buganda Surface backwards a step in time (Bishop, 1965). Nor does this redating of the surface rest only on the evidence provided by the Western Rift Valley. Studies of the volcanics on the eastern boundary of Uganda, Moroto, Napak and Kadam, which began formation in the lower Miocene (Trendall, 1959; Macdonald, 1961; MacGregor, 1962; Bishop 1958, 1962a,b, 1963, 1966; Bishop and Whyte, 1962; Bishop and Trendall, 1967) have shown that they lie in part upon the so-called PIII, having flooded an erosional scarp between this surface and a higher surface. Again the evidence points towards the extensive development of PIII *before* the Lower Miocene when volcanic activity began. A much earlier date is thereby implied for the development of the Buganda Surface. This reassessment of the ages of PIII and consequently PII conforms with the observation that the African (Miocene, Mid-Tertiary) surface is characterized by inselbergs (p. 118) and Hepworth (1961) welcomed this move to redate PIII, pointing out that in the neighbouring territory of Tanganyika it has consistently been called Mid-Tertiary.

This redating of the surfaces in Uganda throws doubt upon another popular generalization concerning laterite genesis, that the Tertiary was the period in which laterite development was most favoured in the tropics. In Uganda, although PIII (Mid-Tertiary) carries a laterite cover, these laterites are generally poorer developments than those of the Buganda and Ntenga Surfaces (formerly Buganda Surface in Kyagwe). They are characteristically immature pisolithic laterites (see Plates 10 and 14) as opposed to the well-developed, mature hydrated massive vermiform laterites of the older surfaces (see Plate 8). This places the major laterite formation here as probably Gondwana or post-Gondwana i.e. Jurassic or Cretaceous. This accords with Dixey's observation that, considering the magnitude of the post-Karoo movements, there is a remarkable absence of terrestrial deposits of Jurassic age. He concluded that the post-Karoo uplift must have been slow enough for the complete breakdown and washing away of the sediments. This in fact postulates an extended period of low relief in which chemical weathering effectively outpaces mechanical erosion — conditions ideal for the development of a residual lateritic crust.

The use of the laterite-capped Buganda Surface as a datum has created many problems and it has effectively lost its significance in this respect. The Western Rift Valley now provides this datum, and the "Buganda Surface" remains problematic, in its identity, mode of development and relationship to other surfaces. If it is to be equated with the Gondwana Surface as King (1951) at first believed and as Ollier (1959) maintained, what are the PI remnants? In the west, the type area of PI, tectonic activity has complicated landsurface analysis, and the nature of the laterite there has never been systematically investigated. Laterite has indeed played a very varied part in the development of theories of landscape evolution in Uganda. With a closer study of the relationship of laterite to landsurface, and in particular of laterite type to landsurface type, laterite should make a clear and positive contribution to our general understanding of landsurface evolution in the tropics.

References

Ab'Saber, A. N. (1959). Conhecimentos sôbre as fluctuações climaticas do quaternário no Brasil. *Noticia geomorfologica* 1(1).

Alexander, L. T. and Cady, J. G. (1962). Genesis and hardening of laterite in soils. *U.S. Dep. Agric. Soil Conservation Service Tech. Bull.* No. 1282.

Alley, N. F. (1970). Duricrust — an Australian dilemma. Paper presented to 1970 meeting of Association of Canadian Geographers.

American Geological Institute. (1962). "Dictionary of Geological Terms". New York.

Andrew, G. (1954). Iron ores in the Anglo-Egyptian Sudan. *In* "Symposium sur les Gisements de fer du Monde". (Blondel, F. and Marvier, L., Eds) *C.R. 19th Int. Geol. Congr.*, Algiers, 1, 187—189.

Babington, B. (1821). Remarks on the geology of the country between Tellicherry and Madras. *Trans. Geol. Soc. Lond.* 5, 328—339.

Bain, A. G. (1852). On the geology of South Africa. *Trans. Geol. Soc.* 7, 175—192.

Baldwin, M., Kellogg, C. E. and Thorpe, J. (1938). "Soil Classification. Soils and Man", 979—1001. Yearbook of Agriculture for 1938 of USDA.

Barshad, I. (1951). Factors affecting the molybdenum content of plants. *Soil Sci.* 71, 297—313.

Barshad, I. and Rojas-Cruz, L. A. (1950). A pedological study of a podzol soil profile from the equatorial region of Colombia, S. America. *Soil Sci.* 70, 221—236.

Bauer, M. (1898). Beiträge zur Geologie der Seychellen insbesondere zur Kenntnis des Laterits. *Neues Jb. Miner. Geol. Palaont.* 2, 163—219.

Baver, L. D. (1956). "Soil Physics" (3rd edn) New York.

Beater, B. E. (1940). Concretions and refractory deposits in some Natal coastal soils. *Soil Sci.* 50, 313—329.

Beckwith, R. S. and Reeve, R. (1964). II: Studies of soluble silica in soils. *Aust. J. Soil Res.* 2, 33—45.

Benza, P. M. (1836). Memoir of the geology of the Neelgherry and Koondah Mountains. *Madras J. Lit. Sci.* 4(13), 241—299.

Bishop, W. W. (1958). Miocene mammalia from the Napak volcanics, Karamoja, Uganda. *Nature, Lond.* 182, 1480—1482.

Bishop, W. W. (1962a). Tertiary mammalian faunas and sediments in Karamoja and Kavirondo, East Africa. *Nature, Lond.* 196, 1283—1287.

Bishop, W. W. (1962b). The mammalian fauna and geomorphological relations of the Napak volcanics, Karamoja. *Rec. geol. Surv. Uganda, 1957—8.* 1—18.

REFERENCES

Bishop, W. W. (1963). The later Tertiary and Pleistocene in eastern Equatorial Africa. *In* "African Ecology and Human Evolution", (Howell, F. C. and Bouliere, F., Eds), 246—275. Aldine, Chicago.

Bishop, W. W. (1965). Quarternary geology and geomorphology in the Albertine Rift Valley, Uganda. *Spec. Pap. geol. Soc. Am.* **84**, 293—321.

Bishop, W. W. (1966). Stratigraphical geomorphology. *In* "Essays in Geomorphology" (Dury, G. H., Ed.), 139—176. Heinemann, London.

Bishop, W. W. (1967). The Lake Albert Basin. *Geogrl. J.* 133(4), 469—480.

Bishop, W. W. (1968). The evolution of fossil environments in East Africa. *Trans. Leic. lit. phil. Soc.* **62**, 22—44.

Bishop, W. W. and Trendall, A. F. (1967). Erosion-surfaces, tectonics and volcanic activity in Uganda. *Q. Jl geol. Soc. Lond.* **122**, 385—420.

Bishop, W. W. and Whyte, F. (1962). Tertiary mammalian faunas and sediments. in Karamoja and Kavirondo, East Africa. *Nature, Lond.* **196**, 1283—1287.

Bishopp, D. W. (1937). The formation of laterite. *Geol. Mag.* **74**, 442—444.

Bisset, C. B. (1937). Hill-top hollows in Masaka District. *Uganda J.* **5**, 130—133.

Blanford, W. T. (1859). Notes on the laterite of Orissa. *Mem. geol. Surv. India* **1**, 280—294.

Blondel, F. (1954). Les gisements de fer de l'Afrique Occidentale Francaise. *In* Symposium sur les gisements de fer du monde (Blondel, F. and Marvier, L., Eds). *C. R. 19th Int. Geol. Congr.* Algiers, **1**, 5—34.

Bloomfield, C. (1953). Sesquioxide immobilisation and clay movements in podzolised soils. *Nature, Lond.* **172**, 958.

du Bois, C. G. B. and Jeffery, P. G. (1955). Laterites on Entebbe peninsula. *Colon. Geol. Miner. Resour.* 5(4), 387—408.

Bonifas, M. (1959). *Mém. serv. carte géol. Alsace et Lorraine* No. 17.

Britton, H. T. S. (1925). Electrometric studies of the precipitation of hydroxides: I and II. *J. Chem. Soc.* **127**, 2110—2141.

Bromfield, S. M. (1954). The reduction of iron oxide by bacteria. *J. Soil Sci.* **5**, 129—139.

Bromfield, S. M. (1965). Studies in the relative importance of iron and aluminium in the sorption of phosphate by some Australian soils. *Aust. J. Soil Res.* **3**, 31—44.

Brosh, A. (1970). Observations on the geomorphic relationships of laterite in southeastern Ankole (Uganda). *Jerusalem Studies in Geography* **1**, 153—179.

Brown, E. H. (1961). Britain and Appalachia: a study in the correlation and dating of planation surfaces. *Trans. Inst. Br. Geogr.* **29**, 91—100.

Brown, E. H. (1968). Unpublished report to RS/RGS Brazil Expedition Committee.

Bryan, W. H. (1952). Soil nodules and their significance. *Sir Douglas Mawson Anniversary Volume, University of Adelaide*, 43—53.

Buchanan, F. (1807). "A Journey from Madras through the Countries of Mysore, Kanara and Malabar". Vol. 2, 436—461, 559; Vol. 3, 66, 89, 251, 258, 378. East India Co., London.

Burridge, J. C. and Ahn, P. M. (1965). A spectrographic survey of representative Ghana forest soils. *J. Soil Sci.* 16(2), 296—309.

Calton, W. E. (1959). Generalisations on some Tanganyika soil data. *J. Soil Sci.* 10(2), 169—176.

Campbell, J. M. (1917). Laterite. *Min. Mag. (Lond.)* **17**, 67—77, 120—128, 171—179, 220—229.

Challinor, J. (1961). "A Dictionary of Geology". University of Wales Press, Cardiff.

Chaplin, J. H. and McFarlane, M. J. (1969). The Moniko petroglyphs. *Uganda J.* 31(2), 207—209.

Chevalier, A. (1949). Points de vue nouveaux sur les sols d'Afrique tropicale sur leur dégradation et leur conservation. *Bull. agric. Congo belge.* 40, 1057—1092.

de Chételat, E. (1938). Le modelé latérique de la Guinée française. *Revue. Géogr. phys. Géol. dyn.* 11(1), 5—120.

de Chételat, E. (1947). La genèse et l'évolution des gisements de nickel de la Nouvelle — Calédonie. *Bull. Soc. geol. Fr.* 17, 105—160.

Clare, K. E. (1960). Roadmaking gravels and soils in Central Africa. *Rd. Res. overseas Bull.*, 12.

Clark, J. (1838). VI. On the lateritic formation. *Madras J. Lit. Sci.* 8, 334—346.

Combe, A. D. (1932). The geology of south-west Ankole and adjacent territories with specific reference to the tin deposits. *Mem. geol. Surv. Uganda* 2.

Conniah, T. H. and Hubble, G. D. (1960). Laterites in Queensland. *J. Geol. Soc. Aust.* 7, 373—386.

Cooke, H. B. S. (1957). Observations relating to Quarternary environments in East and southern Africa. *Trans. geol. Soc. S. Afr.* Annexure to 60, 73 pp.

Cornwall, I. W. (1958). "Soils for the Archaeologist". Phoenix House, London. 230 pp.

d'Costa, V., Ghor Obada, W., Hinga, G. and Makin, J. (1966). The soils of the country around Mumias. Appendix to Kenya Govt. Report on sugar potential, Soil Survey Unit, Kenya.

Craig, D. C. and Loughnan, F. C. (1964). Chemical and mineralogical transformations accompanying the weathering of basic volcanic rocks from N.S. Wales. *Aust. J. Soil Res.* 2, 218—234.

Crook, T. (1909). On the use of the term laterite. *Geol. Mag.* 46, 524—526.

Crook, T. (1910). The use of the terms laterite and bauxite. *Geol. Mag.* 47, 233.

Crowther, E. H. (1930). The relationship of climatic and geological factors to the composition of soil clay and the distribution of soil types. *Proc. R. Soc.* 107B, 1—30.

Curtis, C. D. (1970). Differences between lateritic and podzolic weathering. *Geochim. cosmochim. Acta* 34, 1351—1353.

Darwin, C. R. (1844). "Geological Observations on the Volcanic Islands and Parts of South America Visited During the Voyage of H.M.S. Beagle". Smith, Elder and Co., London. 2nd edition, 1876, 647 pp.

Davis, W. M. (1909). "Geographical Essays". Ginn and Co., Boston. Reprinted 1954, 777 pp.

Davis, W. M. (1920). Physiographic relations of laterite. *Geol. Mag.* 57, 429—431.

Deb, B. C. (1949). The movement and precipitation of iron oxides in podzol soils. *J. Soil Sci.* 1, 112—122.

Dey, A. K. (1942). Bauxites and aluminous laterite occurrences in Jashpur State, Eastern States Agency. *Rec. geol. Surv. India* 75, Prof. pap. No. 16.

Dimbleby, G. W. (1961). Transported material in the soil profile. *J. Soil Sci.* 12(1), 12—22.

Dixey, F. (1945). The relation of the main peneplain of central Africa to sediments of lower Miocene age. *Q. Jl geol. Soc. Lond.* 101(3) and (4), 243—253.

REFERENCES

Dixey, F. (1956). Erosion surfaces of Africa; some considerations of age and origin. *Trans. geol. Soc. S. Africa* **59**, 1—16.
Doornkamp, J. C. and Temple, P. H. (1966). Surface, drainage and tectonic instability in part of southern Uganda. *Geogrl J.* **132**(2), 238—252.
Dorman, F. H. and Gill, E. D. (1959). Oxygen isotope palaeotemperature measurements on Australian fossils. *Proc. R. Soc. Vict.* **71**, 73—98.
Doyne, H. C. and Watson, W. A. (1933). Soil formation in southern Nigeria. *J. agric. Sci.* **23**, 208—215.
Dury, G. H. (1969). Rational descriptive classification of duricrusts. *Earth Science Journal* (Waikato) **3**(2), 77—86.
Erhardt, H. (1951). Sur l'importance des phénomènes biologiques dans la formation des cuirasses ferrugineuses en zone tropicale. *C.r. hebd. Séanc. Acad. Sci., Paris* **233**, 805—806.
Evans, J. W. (1910a). The meaning of the term "laterite". *Geol. Mag.* **47**, 189—190.
Evans, J. W. (1910b). The term "laterite". *Geol. Mag.* **47**, 381—382.
Falconer, J. D. (1911). "The Geology and Geography of Northern Nigeria". Macmillan, London. 295 pp.
Faniran, A. (1970a). Maghemite in the Sydney duricrusts. *Am. Miner.* **55**, 925—933.
Faniran, A. (1970b). Landform examples from Nigeria No. 2: the deep-weathering (duricrust) profile. *Nigerian Geog. J.* **13**(1), 87—88.
Faniran, A. (1971). The parent material of Sydney laterites. *J. Geol. Soc. Aust.* **18**(2), 159—164.
Fermor, L. L. (1911). What is laterite? *Geol. Mag.* **48**, 454—462, 507—516, 559—566.
Fisher, N. H. (1958). Notes on lateritisation and mineral deposits. The Australian Institute of Mining and Metallurgy: F. L. Stillwell Anniversary Volume, 133—142.
Follet, E. A. C. (1965). The retention of amorphous colloidal "ferric hydroxide" by kaolinite. *J. Soil Sci.* **16**(2), 334—341.
Foote, R. B. (1880). II. Notes on the occurrence of stone implements in the coast laterite, south of Madras, and in high-level gravels and other formations in the south Mahratta country. *Geol. Mag.* **17**, 542—546.
Fox, C. S. (1927). "Bauxite and Aluminous Laterite". Crosby, Lockwood and Son, London. 312 pp.
Fox, C. S. (1933). Laterite. *Geol. Mag.* **70**, 558.
Frasché, D. F. (1941). Origin of the Surigao iron ores. *Econ. Geol.* **36**, 280—305.
Fripait, J. J. and Gastuche, M. C. (1952). Étude physico-chimique des surfaces des argiles. *Publs Inst. natn. Étude agron. Congo belge*, Sér. Sci. 54.
Gammon, N. Jr., Volk, G. M., McCubbin, E. N. and Eddins, A. H. (1954). Soil factors affecting molybdenum uptake by cauliflower. *Proc. Soil Sci. Soc. Am.* **18**, 302—305.
Gautier, A. (1965). Relative dating of peneplains and sediments in the Lake Albert rift area. *Am. J. Sci.* **263**(6), 537—547.
Gautier, A. (1966). Geschiedenis en evolutie van de zoetwater-molluskenfauna in de Albert-en Edwardmeren-Slenk. *Natuurw. Tijdschr.* (Ghent) **48**, 3—24.
Gautier, A. (1967). New observations on the later Tertiary and early Quarternary in the Western Rift; the stratigraphic and palaeontological

evidence. *In* "Background to Evolution in Africa". (Bishop, W. W. and Clark, J. D., Eds), 73—88. University of Chicago Press, Chicago and London.

Gautier, A. and Geets, S. (1966). Zware mineralen van de zoetwater afzettingen in de Albert-en Edwardmeren-Slenk. *Natuurw. Tijdschr.* (Ghent) **48**, 141—156.

Gear, D. J. (1955). The electrical resistivity problem in the location of groundwater in the decomposed igneous and metamorphic rocks of Uganda and Sudan. Thes. Doct. Univ. London.

Geddes, W. R. (1960). The human background. *In* "Symposium on the Impact of Man on Humid Tropics Vegetation", 42—56. Sponsored by the administration of the Territory of Papua and New Guinea and Unesco Science Co-operation Office for South-East Asia.

Gentilli, J. (1961). Quarternary climates of the Australian region. *Ann. N. Y. Acad. Sci.* **95**, 463—501.

Gill, E. O. (1961). Cainozoic climates of Australia. *Ann. N. Y. Acad. Sci.* **95**, 461—446.

Glinka, K. D. (1914). "Die Typen der Bodenbildung, ihre Klassification und Geographische Verbreitung" (Translated by C. F. Marbut). Borntraeger, Berlin.

Goldman, M. I. and Tracy, J. I. (1946). Relations of bauxite and kaolin in the Arkansas bauxite deposits. *Econ. Geol.* **41**(6), 567—575.

Goudie, A. (1973). "Duricrusts in Tropical and Subtropical Landscapes". Oxford Research Studies in Geography. Oxford University Press, London. 174 pp.

Gozan, B. M. and Vera, E. C. (1962). Aluminous laterite deposits in northern Mindanao. Potential sources of refractory raw material. *Minerals News Serv. Philipp. Isl.* No. 46.

Griffith, G. (1953). Vesicular laterite. *Nature, Lond.* **171**, 530.

Grim, R. E. (1953). "Clay Mineralogy". McGraw Hill, New York. 384 pp.

Grubb, P. L. C. (1963). Critical factors in the genesis, extent and grade of some residual bauxite deposits. *Econ. Geol.* **58**(8), 1267—1277.

Hamming, E. (1968). On laterites and latosols. *Prof. Geogr.* **20**(4), 238—241.

Handley, J. R. F. (1954). The geomorphology of the Nzega area of Tanganyika with special reference to the formation of granite tors. *C.r. 19th Int. Geol. Congr.*, Algiers, Fasc. 21, 201—210.

Hanlon, F. N. (1945). The bauxites of N.S. Wales, their distribution, composition and probable origin. *J. Proc. R. Soc. N.S.W.* **78**, 94—112.

Harden, G. and Bateson, J. H. (1963). A geochemical approach to the problem of bauxite genesis in British Guiana. *Econ. Geol.* **58**, 1301—1308.

Hardy, F. (1931). Studies on tropical soils. I: Identification and approximate estimation of sesquioxide components by absorption of alizarin. *J. Soil Sci.* **21**, 120—166.

Hardy, F. and Follett-Smith, R. R. (1931). Studies on tropical soils. II; Some characteristic igneous rock soil profiles in British Guiana, South America. *J. agric. Sci.* **21**, 239—261.

Harrassowitz, H. (1930). Boden des tropischen Region. Laterit und allitscher (lateritisher) Rotlehm. *In* "Handbuch der Bodenlehre". (Blanck, E., Ed.), Vol. 3, 387—436. Berlin.

Harris, N. (1947). Description of the Kampala Map Sheet. Unpublished Report, Geol. Surv. Uganda, NH/39.

REFERENCES

Harrison, J. B. (1911). IV: On the formation of a laterite from a practically quartz-free diabase. *Geol. Mag.* 48, 120—123, 477—478.
Harrison, J. B. and Reid, K. D. (1910). III, IV and V: The residual earths of British Guiana commonly termed "laterite". *Geol. Mag.* 47, 439—452, 488—495, 553—562.
Hartman, J. A. (1955). Origin of heavy minerals in Jamaican bauxite. *Econ. Geol.* 50(7), 738—747.
Hatton, A. (1953). Report on Sheet South A.36, B.I.S.W., Kabula, Masaka. Unpublished Report, Geol. Surv. Uganda, AH/9.
Hays, J. (1967). Land surfaces and laterites of the Northern Territory. *In* "Landform Studies from Australia and New Guinea". (Jennings, J. N. and Mabbut, J. A., Eds), 182—210. Cambridge.
Heintze, S. G. (1946). Manganese deficiency in peas and other crops in relationship to the availability of soil manganese. *J. agric. Sci.* 36(4), 227—238.
Heintze, S. G. and Mann, P. J. G. (1947). Soluble complexes of manganio manganese. *J. agric. Sci.* 37(1), 23—26.
de Heinzelin, J. (1962). Les formations du Western Rift et de la Cuvette Congolaise. *Proc. pan-Afr. Congr. on Prehistory 1959*, 219—243.
de Heinzelin, J. (1963). Palaeoecological conditions of the Lake Albert-Lake Edward Rift. *In* "African Ecology and Human Evolution". (Howell, F. C. and Bouliere, F., Eds), 276—284. Aldine, Chicago.
Hepworth, J. V. (1951). The laterite between Banda and Nazigo, Kyaggwe. Unpublished Report, Geol. Surv. Uganda, JVH/2.
Hepworth, J. V. (1952a). Laterite profiles in the Banda-Nazigo-Sezibwa area. Unpublished Report, Geol. Surv. Uganda, JVH/4.
Hepworth, J. V. (1952b). A laterite surface in Kyaggwe, Buganda. Unpublished Report, Geol. Surv. Uganda, JVH/5.
Hepworth, J. V. (1961). The geology of southern West Nile, Uganda, with particular reference to the charnockites and to the development of the Albert rift. Thes. Doct. Univ. Leeds, 2.
Holland, T. H. (1903). On the constitution, origin and dehydration of laterite. *Geol. Mag.* 40, 59—69.
Holmes, A. (1914). The lateritic deposits of Mozambique. *Geol. Mag.* 51, 529—537 (decade 6).
Holmes, A. (1945). "Principles of Physical Geology" Nelson, London. 532 pp.
Hooijer, D. A. (1963). Miocene mammalia of Congo. *Annls Mus. r. Afr. cent. (geol.)* 46, 71 pp.
d'Hoore, J. (1954). L'accumulation des sesquioxides libres dans les sols tropicaux. *Publs Inst. natn Étude agron. Congo belge*, Sér. Sci. 62, 132 pp.
Humbert, R. P. (1948). The genesis of laterite. *Soil. Sci.* 65, 281—290.
Imperial Bureau of Soil and Science. (1932). Laterite and laterite soils. *I.B.S.S. Tech. Comm.* No. 24.
Islah, M. A. and Elahi, M. A. (1954). Reversion of ferric iron to ferrous iron under waterlogged conditions and its relation to available phosphorous. *J. agric. Sci.* 45, 1—2.
Jenny, H. (1929). Klima und Klimabodentypen in Europa und in den Vereiningten Staaten von Nordamerika. *Soil Res.* 1, 139—189.
Jensen, H. I. (1911). The nature and origin of Gilgai. *J. Proc. R. Soc. N.S.W.* 45, 337—358.

Jensen, H. I. (1914). Geological Report on the Darwin Mining District; McArthur River District; Barkly Tableland. *Bull. Northern Territory* (Melbourne), No. 10.

Jessup, R. W. (1960). Laterite soils of the Australian arid zone. *J. Soil Sci.* 2(1), 106—113.

Johnson, R. J. (1954). Preliminary report on sheet NA36 T IV SE (Kabulosoke Sheet). Unpublished Report, Geol. Surv. Uganda, RJJ/7.

Johnson, R. J. (1956). Description of the Wamala Sheet. Unpublished Report, Geol. Surv. Uganda, RJJ/13.

Johnson, R. J. (1959). Physiographic evolution of western Buganda District. Unpublished report, Geol. Surv. Uganda, RJJ/23.

Johnson, R. J. (1960). Explanation of the geology of sheet 69 (Lake Wamala). *Rep. geol. Surv. Uganda* 3.

Johnson, R. J. and Williams, C. E. F. (1961). Explanation of the geology of sheet 59 (Kiboga). *Rep. geol. Surv. Uganda* 7.

Jones, L. H. P. (1957). The solubility of molybdenum in simplified systems and aqueous soil suspensions. *J. Soil Sci.* 8(2), 313—327.

Kellog, C. E. (1949). Preliminary suggestions for the classification and nomenclature of great soil groups in tropical and equatorial regions. *Commonw. Bur. Soil Sci. (Gt. Brit.) Tech. Commun.* No. 46, 76—85.

Kellog, C. E. (1962). Introduction to Alexander, L. T. and Cady, J. G. (1962) *op. cit.*

Kent, P. E. (1944). The Miocene of Kavirondo, Kenya. *Q. Jl geol. Soc. Lond.* 100, 85—118.

von Kerner-Marilaun, F. (1927). Die klimatischen Schwellenwert d. voltstaendigen Laterits Profils. *Akad. Wiss. Math.-nat. Kl. Wien*, alt. IIa. 136(7), 413—428.

King, B. C. (1957). The geomorphology of Africa. I: Erosion surfaces and their mode of origin. *Sci. Prog. Lond.* 45, 672—681.

King, L. C. (1951). "South African Scenery: a Textbook of Geomorphology" (2nd edn). Oliver and Boyd, Edinburgh. 379 pp.

King, L. C. (1962). "Morphology of the Earth". Oliver and Boyd, London. 699 pp.

King, W. (1882). *Laterite in Travancore State. Rec. geol. Surv. India* No. 15, 96—97.

Krauskopf, K. B. (1956). Dissolution and precipitation of silica at low temperatures. Silica. *Geochim. cosmochim. Acta.* 10, 1—26.

Krauskopf, K. B. (1967). "Introduction to Geochemistry". McGraw-Hill, New York.

Lacroix, A. (1913). Les latérites de la Guinée et les produits d'alteration qui leur sont associés. *Nouv. Archs. Mus. Hist. nat., Paris* Ser. 5, 5, 255—358.

Lake, P. (1933). Buchanan's laterite. *Geol. Mag.* 70, 240.

Laruelle, J. (1961). Les grandes catenas au Parc National de la Kagera (Ruanda). *Pédologie, Gand* 11(1), 158—216.

Leakey, L. S. B. (1967). Discussion in Gautier, A. (1967) *op. cit.*, 85—87.

Lelong, F. (1966). Régime des nappes phréatiques contenues dans les formations d'altération tropicale. Conséquences pour la pédogenèse. *Sci. Terre* 11, 203—244.

Lepersonne, J. (1956). Les aplanissements d'érosion du nord-est du Congo Belge et des regions voisines. *Mém. Acad. r. Sci. colon.* No. 4, 1—108.

Lindgren, W. (1925). Gel-metasomatism or replacement of crystalloids by gels (Presidential address). *Bull. geol. Soc. Am.* 36, 253—255.

Litchfield, W. H. and Mabbutt, J. A. (1962). Hardpan in soils of semi-arid Western Australia. *J. Soil Sci.* 13, 148—159.

Loughnan, F. C., Grim, R. E. and Vernet, J. (1962). Weathering of some Triassic shales in the Sydney area. *J. geol. Soc. Aust.*, Adelaide, 8(2), 245—257.

Mabbutt, J. A. (1961). A stripped land surface in western Australia. *Trans. Inst. Br. Geogr.* 29, 101—114.

Macdonald, R. (1961). Explanation of the geology of sheet 36 (Nabilatuk). *Rep. geol. Surv. Uganda* 5.

MacGregor, P. (1962). Explanation of the geology of sheet 10 (Kaabong). *Rep. geol. Surv. Uganda* 9.

Maclaren, M. (1906). On the origin of certain laterites. *Geol. Mag.* 43, 536—547.

Maignien, R. (1958). Le cuirassement des sols en Guinée Afrique Occidental. *Mém. Serv. carte géol. Alsace et Lorraine* 16.

Maignien, R. (1966). "Review of Research on Laterites". Unesco, Natural Resources Research Se., IV., Paris.

Marbut, C. F. and Manifold, C. B. (1926). The soils of the Amazon basin in relation to agricultural possibilities. *Geogrl Rev.* 16, 414—442.

Martin, F. J. and Doyne, H. C. (1927). I: Laterite and lateritic soils in Sierra Leone. *J. agric. Sci.* 17, 530—547.

Martin, F. J. and Doyne, H. C. (1930). II: Laterite and lateritic soils in Sierra Leone. *J. agric. Sci.* 20, 135—143.

Mattson, S. (1941). The laws of colloidal behaviour: xxiii. The constitution of the pedosphere and soil classification. *Soil Sci.* 51, 407—425.

Maud, R. R. (1965). Laterite and lateritic soils in coastal Natal, South Africa. *J. Soil Sci.* 16(1), 60—72.

Maufe, H. B. (1933). Laterite. *Geol. Mag.* 70, 144.

McConnell, R. B. (1955). The erosion surfaces of Uganda. *Colon. Geol. Miner. Resour.* 5, 425—428.

McFarlane, M. J. (1968). Some observations on the prehistory of the Buvuma Island group of Lake Victoria. *Rep. E. Afr. Freshwat. Fish Res. Org.* 1967, Appendix f, 49—54.

McFarlane, M. J. (1969). Lateritisation and landscape development in parts of Uganda. *Thes. Doct. Univ. London.*

McFarlane, M. J. (1971). Lateritization and landscape development in Kyagwe, Uganda. *Q. Jl Geol. Soc. Lond.* 126, 501—539.

McFarlane, M. J. (1973). Laterite and topography in Buganda. *Uganda J.* 36 (for 1972), 9—22.

McFarlane, M. J. (1976). A calcrete from the Namanga-Bissel area of Kenya. *The Kenyan Geographer* 2(1).

McFarlane, M. J. (in press). A geomorphic classification of laterites in Uganda.

McGee, W. J. (1880). The "laterite" of the Indian Peninsula. *Geol. Mag.* 17, 310—313.

McNeil, M. (1964). Lateritic soils. *Sci. Am.* 211(5), 96—102.

Mennell, F. P. (1909). IV; Notes on Rhodesian laterite. *Geol. Mag.* 46, 350—352.

Mikhaylov, B. M. (1964). Part played by vegetation in the lateritization of mountainous areas of the Liberian Shield. *Dokl. Akad. Nauk SSSR* 157, 23—24.

Mitsuchi, T. (1954). Iron ore deposits in Japan. *In* "Symposium sur les Gisements de Fer du Monde". (Blondel, F., and Marvier, L., Eds) *C.r. 19th Int. Geol. Congr.*, Algiers, **1**, 537—560.

Mohr, E. C. J. (1944). "Soils of Equatorial Regions". (transl. by R. L. Pendelton) , Edwards, Ann Arbor, Michigan.

Mohr, E. C. J. and Van Baren, F. A. (1954). "Tropical Soils" Interscience, New York. 498 pp.

Moore, E. S. and Maynard, J. E. (1929). Solution, transportation and precipitation of iron and silica. *Econ. Geol.* **24**, 272—303, 365—402, 506—527.

Moorman, F. R. and Panabokke, C. R. (1961). Soils of Ceylon. *Trop. Agric. Mag. Ceyl. agric. Soc.* 117, 5—65.

Moss, R. P. (1965). Slope development and soil morphology in a part of south-west Nigeria. *J. Soil Sci.* 16(2), 192—209.

Muir, A., Anderson, B. and Stephens, I. (1957). Characteristics of some Tanganyika soils. *J. Soil Sci.* 8(1), 1—18.

Mulcahy, M. J. (1960). Laterites and lateritic soils in south-western Australia. *J. Soil Sci.* 11, 206—226.

Mulcahy, M. J. (1961). Soil distribution in relation to landscape development. *Z. Geomorph.* (NS) 5, 211—225.

Mulcahy, M. J. and Hingstone, F. J. (1961). The development and distribution of the soils of the York-Quairading area, Western Australia, in relation to landscape evolution. *Soil Publ. C.S.I.R.O. Aust.* No. 17.

Nagell, R. H. (1962). Geology of the Serra do Navio Manganese District, Brazil. *Econ. Geol.* 57(4), 481—498.

Nazaroff, P. S. (1931). Notes on the spongy ironstone of Angola. *Geol. Mag.* **68**, 443—446.

Netterberg, F. (1966). Calcrete Research Project Progress Report, RS/8/7 CSIR, NIRR, Pretoria.

Newbold, T. J. (1844). Notes, chiefly geological, across the Peninsula from Masulipatam to Goa, comprising remarks on the origin of the regur and laterite: occurrence of manganese veins in the latter and on certain traces of aqueous denudation on the surface of southern India. *J. Asiat. Soc. Beng.* **13**, 984—1004.

Newbold, T. J. (1846a). Notes chiefly geological on the western coast of South India. *J. Asiat. Soc. Beng.* **15**, 204—213, 224—231, 380—396.

Newbold, T. J. (1846b). Summary of the geology of S. India. VI: Laterite. *R. Asiat. Soc.* **8**, 227—240.

Nye, P. H. (1954). Some soil-forming processes in the humid tropics. I: A field study of a catena in the West African forest. *J. Soil Sci.* 5(1), 7—21.

Nye, P. H. (1955). Some soil-forming processes in the humid tropics. II: The development of the upper slope member of the catena. III: Laboratory studies of the development of a typical catena over granitic gneiss. *J. Soil Sci.* 6(1), 51—72.

Oakley, K. P. (1961). "Man the Toolmaker". (5th ed., reprinted 1965) British Museum, Natural History, London. 98 pp.

Oertel, G. (1956). Studie über einen Laterit in Goa. *Neues Jb. Geol. Paläont. Abh.*, Stuttgart, 104(2), 148—180.

Oldham, R. D. (1893). "A Manual of the Geology of India". (2nd edn), Calcutta, 369—390.

REFERENCES

Ollier, C. D. (1959). A two-cycle theory of tropical pedology. *J. Soil Sci.* 10(2), 137–148.
Ollier, C. D. (1960). The inselbergs of Uganda. *Z. Geomorph.* (NS) 4(1), 43–52.
Page, D. (1859). "Handbook of Geological Terms". Blackwood, Edinburgh.
Pallister, J. W. (1951). Occurrence of laterite in South Buganda. Unpublished Report, Geol. Surv. Uganda, JWP/7.
Pallister, J. W. (1953). Notes on the geomorphology of the southern part of Mawokota and Busiro Counties (sheet U. III. SE). Unpublished Report, Geol. Surv. Uganda, JWP/21.
Pallister, J. W. (1954). Erosion levels and laterite in Buganda Province, Uganda. *C.r. 19th Int. Geol. Congr.*, Algiers, fasc. 21, 193–199.
Pallister, J. W. (1955). The physiography of south central Uganda. Unpublished Report, Geol. Surv. Uganda, JWP/28.
Pallister, J. W. (1956a). The physiography of Mmengo District Buganda. Unpublished Report, Geol. Surv. Uganda, JWP/28a.
Pallister, J. W. (1956b). Slope form and erosion surfaces. Unpublished Report, Geol. Surv. Uganda, JWP/32.
Pallister, J. W. (1956c). Slope development in Buganda. *Geogrl J.* 122, 80–87.
Pallister, J. W. (1956d). Slope form and erosion surfaces. *Geol. Mag.* 93, 465–472.
Pallister, J. W. (1957). The physiography of Mengo District, Buganda. *Uganda J.* 21, 16–29.
Pallister, J. W. (1959). The geology of southern Mengo. *Rep. geol. Surv. Uganda* 1.
Pallister, J. W. (1960). Erosion cycles and associated surfaces of Mengo District, Buganda. *Overseas Geol. Miner. Resour.* 8, 26–36.
Panton, W. P. (1956). Types of Malayan laterite and factors affecting their distribution. *Report, 6th Int. Congr. Soil Sci.*, Paris, 5(69), 419–423.
Patz, M. J. (1965). Hill-top hollows – further investigations. *Uganda J.* 29, 225–228.
Penck, W. (1953). "Morphological Analysis of Land-forms" (Trans. by H. Czech and K. C. Boswell, of *Die Morphologische Analyse*). Macmillan, London. 429 pp.
Pendleton, R. L. (1941). Laterite and its structural uses in Thailand and Cambodia. *Geogrl Rev.* 31(2), 177–202.
Pendleton, R. L. and Sharasuvana, S. (1946). Analyses of some Siamese laterites. *Soil Sci.* 62, 423–440.
Pickering, R. J. (1962). Some leaching experiments on three quartz-free silicate rocks and their contribution to an understanding of laterization. *Econ. Geol.* 57, 1185–1206.
Piper, C. S. (1931). The availability of manganese in the soil. *J. agric. Sci.* 21, 762–779.
Playford, P. E. (1954). Observations of laterite in Western Australia. *Aust. J. Sci.* 17(1), 11–14.
Plummer, H. G. (1960). Explanation of the geology of sheet 86. Unpublished manuscript, Geol. Surv. Uganda.
du Preez, J. W. (1949). Laterite. A general discussion with a description of Nigerian occurrences. *Bull. agr. Congo belge* 40, 53–66.
du Preez, J. W. (1954). Notes on the occurrence of oolites and pisolites in Nigerian laterites. *C.r. 19th Int. Geol. Congr.*, Algiers, fasc. 21, (1952), 163–169.

Prescott, J. A. and Pendleton, R. L. (1952). Laterite and lateritic soils. *Commonw. Bur. Soil Sci. Tech. Commun.* No. 47.

Proudfoot, V. B. (1967). Experiments in archaeology. *Science Journal* (IPC Business Press Ltd., Lond.) 3, (II), 59—64.

Pulfrey, W. (1960). Shape of the sub-Miocene erosion bevel in Kenya. *Bull. geol. Surv. Kenya* 3.

Pullan, R. A. (1967). A morphological classification of lateritic ironstones and ferruginised rocks in Northern Nigeria. *Nigerian Journal of Science* 1(2), 161—174.

Radwansky, S. A. and Ollier, C. D. (1959). A study of an East African capital Catena. *J. Soil Sci.* 10(2), 149—168.

Rao, T. V. M. (1928). A study of bauxite. *Min. Mag.* 21, 407—430.

Read, H. M. (1947). "Rutley's Elements of Mineralogy". (24th edn) Murby, London. 525 pp.

Reiche, P. (1950). A survey of weathering processes and products. *Univ. New Mex. Publs Geol.* No. 1.

Reifenberg, A. (1935). Soil formation in the Mediterranean. *3rd Int. Congr. Soil Sci.* 6, 306—309.

Reifenberg, A. (1938). "The Soils of Palestine: Studies in Soil Formation and Land Utilisation in the Mediterranean". Murby, London. 131 pp.

Reisenauer, H. M., Tabikh, A. A. and Stout, P. R. (1962). Molybdenum reactions with soils and the hydrous oxides of iron, aluminium and titanium. *Proc. Soil Sci. Soc. Am.* 26, 23—27.

Rice, C. M. (1957). "Dictionary of Geological Terms". Edwards, Ann Arbor.

Robinson, W. O. and Holmes, R. S. (1924). The chemical composition of soil colloids. *Bull. U.S. Dep. Agric.* No. 1311.

Rodin, L. E. and Bazilevich, N. I. (1967). "Production and Mineral Cycling in Terrestrial Vegetation". Edinburgh.

Roy Chowdhury, M. K., Anandalwar, M. A. and Paul, D. K. (1965). Recent concepts of the origin of Indian laterite. *Proc. natn. Inst. Sci. India* Part A, 31(6), 547—558.

Ruhé, R. V. (1954). Erosion surfaces of Central African interior high plateaus. *Publs Inst. natn Étude agron. Congo belge* No. 59, 1—40.

Ruhé, R. V. (1956). Landscape evolution in the High Ituri. *Inst. Natn Etude agron. Congo belge*, Sér. Sci. No. 66, 91 pp.

Russell, E. W. (1962). "Soil Conditions and Plant Growth" (9th ed., new impression). Longmans, London. 688 pp.

Russell, I. C. (1889). Sub-aerial decay of rocks and origin of red colour in certain formations. *U.S. Geol. Soc. Bull.* No. 52.

Sabot, J. (1954). Les laterites. *C.r. 19th Int. Geol. Congr.*, Algiers, fasc. 21, 181—192.

Saggerson, E. P. and Baker, B. H. (1965). Post-Jurassic erosion surfaces in eastern Kenya and their deformation in relation to rift structures. *Q. Jl geol. Soc. Lond.* 121, 51—72.

Sandford, K. S. (1935). Geological observations on the north-west Frontiers of the Anglo-Egyptian Sudan and the adjoining part of the South Libyan Desert. *Q. Jl geol. Soc. Lond.* 91, 323—381.

Sato, M. (1960). Oxidation of sulfide ore bodies. 1: Geochemical environments in terms of Eh and pH. *Econ. Geol.* 55, 928—961.

REFERENCES

Schieferdecker, A. A. G. (Ed.). (1959). "Geological Nomenclature" Roy. Geol. Mining Soc. Netherlands. Garinchen.

Schnell, R. (1949). Observations sur l'instabilité de certaines forêts de la Haute-Guinée francaise en rapport avec le modelé et la nature du sol. *Bull. agric. Congo belge* 40(1), 671–676.

Schnitzer, M. and Skinner, S. I. M. (1964). Organo-metallic interactions in soils. *Soil Sci.* 98(3), 197–203.

Scrivenor, J. B. (1909). The use of the word "laterite". *Geol. Mag.* 46, 431–432, 574–575.

Scrivenor, J. B. (1910a). The use of the term laterite. *Geol. Mag.* 47, 139.

Scrivenor, J. B. (1910b). The term laterite. *Geol. Mag.* 47, 335.

Scrivenor, J. B. (1910c). Laterite and bauxite. *Geol. Mag.* 47, 382.

Scrivenor, J. B. (1932). Review of "Bauxite and Aluminous Laterite", Fox, C. E. (Crosby, Lockwood & Son, London), 312 pp., *Geol. Mag.* 69, 559–560.

Scrivenor, J. B. (1933). Laterite. *Geol. Mag.* 70, 191.

Scrivenor, J. B. (1937). Note on Buchanan's laterite. *Geol. Mag.* 47, 256–262.

Shackleton, R. M. (1951). A contribution to the geology of the Kavirondo Rift Valley. *Q. Jl geol. Soc. Lond.* 106, 345–392.

Simmons, W. C. (1929). Lateritisation and peneplanation. *Ann. Rep., Geol. Surv. Uganda*, 1928, 41.

Simpson, E. S. (1912). Notes on laterite in Western Australia. *Geol. Mag.* 49, 399–406.

Sivarajasingham, S., Alexander, L. T., Cady, J. G. and Cline, M. G. (1962). Laterite. *Adv. Agron.*, New York 14, 1–60.

Sombroek, W. G. (1966). "Amazon Soils". Centre for Agricultural Publications and documentation, Wageningen.

Sombroek, W. G. (1971). Ancient levels of plinthisation in N.W. Nigeria. *In* "Paleopedology: Origin, Nature and Dating of Paleosols". (Yaalon, D. H., Ed.), 329–336. Jerusalem.

Stamp, L. D. (Ed.) (1961). "A Glossary of Geographical Terms". Longman, London. 539 pp.

Stephens, C. G. (1961). Laterite in the type locality, Angadipuram, Kerala, India. *J. Soil Sci.* 12(2), 214–217.

Stockley, G. M. and Williams, G. J. (1938). Explanation of the Geology Degree Sheet No. 1 (Karagwe tinfields). *Bull. Geol. Surv. Tanganyika* 10.

de Swardt, A. M. J. (1964). Lateritisation and landscape development in parts of Equatorial Africa. *Z. Geomorph.* (NS) 8(3), 313–333.

de Swardt, A. M. J. and Trendall, A. F. (1970). The physiographic development of Uganda. *Overseas Geol. Miner. Resour.* 10, 241–288.

Taylor, R. M. and McKenzie, R. M. (1964). The mineralogy and chemistry of manganese in some Australian soils. *Aust. J. Soil Res.* 2, 235–248.

Teixeira, C. (1965). Les Latérites de Goa et le problème de leur genèse. *Garcia de Orta* (Lisboa) 13(1), 69–86.

Tessier, F. (1959). La latérite du Cap Manuel à Dakar et ses termitières fossiles. *C.r. Acad. Sci., Paris,* 248, 3320–3322.

Thiel, G. A. (1925). Manganese precipitation by micro-organisms. *Econ. Geol.* 20(4), 301–310.

Thomas, M. F. (1968). Some outstanding problems in the interpretation of the geomorphology of tropical shields. British Geomorphological Research Group, Occasional Paper. 5, 41–49.

Thomas, M. F. (1974). "Tropical Geomorphology". Focal problems in geography, Macmillan, London. 332 pp.

Thorpe, J. and Baldwin, M. (1940). Laterite in relation to soils of the tropics. *Ann. Ass. Am. Geogr.* 30, 163—194.

Tiller, K. G. (1963). Weathering and soil formation on dolerite in Tasmania with particular reference to several trace elements. *Aust. J. Soil Res.* 1, 74—90.

Trendall, A. F. (1959). The topography under the northern part of Kadam volcanics and its bearing on the correlation of the peneplains of south-east Uganda and the adjacent parts of Kenya. *Rec. geol. Surv. Uganda* 1955—6, 1—8.

Trendall, A. F. (1962). The formation of apparent peneplains by a process of combined lateritisation and surface wash. *Z. Geomorph.* (NS) 6(2), 183—197.

U.S. Soil Conservation Service. (Soil Survey Staff). (1960). Soil classification. A comprehensive system. 7th Approximation. *Publs U.S. Dep. Agric.*

Van Bemmelen, R. W. (1941). Origin and mining of bauxite in Netherlands India. *Econ. Geol.* 36, 630—640.

Van der Eyk, J. J. (1965). Climate as a soil forming factor in Natal. *Proc. S. Afr. Sug. Technol. Ass.* 1965, 1—8.

Van der Merwe, C. R. and Heystek. H. (1952). Clay minerals of South African soil groups. I: Laterites and related soils. *Soil Sci.* 74, 383—401.

Vann, J. H. (1963). Developmental processes in laterite terrain in Amapa. *Geog. Rev.* 53(3), 406—417.

Vilensky, D. G. (1925). The classification of soils on the basis of analogous series in soil formation. *Proc. Int. Soc. Soil Sci.* 1, 224—241.

de Vletter, D. R. (1955). How Cuban nickel ore was formed — a lesson in laterite genesis. *Engng Min. J.* 156(10), 84—87, 178.

Von Schellmann, W. (1964). Zur lateritischen Verwitterung von serpentinit. *Geol. Jber. Hannover* 81, 645—678.

Voysey, H. H. (1833). Report on the geology of Hyderabad. *J. Asiat. Soc. Beng.* 2, 298—305, 392—405.

Walther, J. (1915). Laterit in Westaustralien. *Z. dt. geol. Ges.* 67B(4), 113—140.

Walther, J. (1916). Das geologische Alter und die Bildung des Laterits. *Petermanns geogr. Mitt.* 62, 1—7, 46—53.

Warth, H. and Warth, F. J. (1903). The composition of Indian laterite. *Geol. Mag.* 40, 154—159.

Watson, J. P. (1965). A soil catena on granite in southern Rhodesia. *J. Soil Sci.* 16(1), 158—169.

Wayland, E. J. (1921). A general account of the geology of Uganda by the geologist. *Rep. geol. Dep. Uganda.* 1921, 8—20.

Wayland, E. J. (1931). Summary of progress of the Geological Survey of Uganda. *Summ. Prog. geol. Surv. Uganda* 1919—29.

Wayland, E. J. (1932). Notes on Kololo Hill — Kampala. Unpublished Report, Geol. Surv. Uganda.

Wayland, E. J. (1933). The peneplains of East Africa. *Geogrl J.* 82, 95.

Wayland, E. J. (1934a). Peneplains and some other erosional platforms. Annual Report, Geol. Surv. Uganda, 1933, 77—78.

Wayland, E. J. (1934b). The peneplains of East Africa. *Geogrl J.* 83, 79.

Wayland, E. J. (1935). Short account of the geology of Ankole District. Unpublished manuscript, Geol. Surv. Uganda.

REFERENCES

Webster, R. (1965). A catena of soils on the Northern Rhodesia Plateau. *J. Soil Sci.* 16(1), 31—43.

Weiss, F. (1910). Verkommen und Entstehung der Kaolin erden des astthuringishen Buntsand steinbeckens. *Z. prakt. Geol.* 18, 353—367.

de Weiss, G. (1954). Note sur quelques types de laterite de la Guinee Portugaise. *C.r. 19th Int. Geol. Congr.*, Algiers, fasc. 21, 171—179.

Wells, N. (1956). Soil studies using sweet vernal for assessing element availability. *N.Z. Jl Sci. Technol.* 37, 482—502.

Wilhelmy, H. (1952). Die eiszeitliche und nacheiszeitliche Verschiebung der Klima-und Vegetationszonen in Südamerika. *Tägl. Ber. u. Wiss. Abh. d. Dt. Geogr. Tage*, Frankfurt am Main, 1951, 121—128.

Willis, B. (1933). The peneplains of East Africa. *Geogrl J.* 82, 383—384.

Willis, B. (1936). "East African Plateaus and Rift Valleys". Studies in comparative seismology, Washington, 358 pp.

Wolfenden, E. B. (1961). Bauxite in Sarawak. *Econ. Geol.* 56, 972—981.

Wood, T. W. W. and Beckett, P. H. T. (1961). Some Sarawak soils. II: Soils of the Bintulu coastal area. *J. Soil Sci.* 12(2), 218—233.

Woolnough, W. G. (1918). The physiographic significance of laterite in Western Australia. *Geol. Mag.* 5, 385—393 (Decade 6).

Woolnough, W. G. (1927). The chemical criteria of peneplanation. (Presidential Address, I); also, The duricrust of Australia (II). *J. Proc. R. Soc. N.S.W.* 61, 1—53.

Young, A. and Stephens, I. (1965). Rock weathering and soil formation on high altitude plateaux of Malawi. *J. Soil Sci.* 16(2), 322—333.

Young, R. S. (1955). Solubility of cobalt in soil. *J. Soil Sci.* 6(2), 233—240.

Subject Index

A

Ability to harden (*see* Induration)
Absolute accumulation (*see also* Detrital laterite; Lateral movement of iron), 95
Accumulation of residual material (*see* Residual laterite)
Acid rocks (*see individual rock types*), 24, 25, 26
Acidity, 49, 81–85, 88, 90
Acylic development of laterite (*see* Pedogenetic laterite; *see also* Mesas, acyclic development)
Adsorption of molybdenum by aluminosilicates, 88
  of phosphates by aluminum, 89
  of phosphates by iron, 89
Aeration (*see* Exposure to air)
African cycle, 109, 110
Albert – Semliki – Edward Rift Valley (*see* Western Rift Valley)
Albertine Rift (*see* Western Rift Valley)
African Surface (*see* Diagnostic characteristics of African Surface; *see also* Mid-Tertiary Surface; Miocene Surfaces; Sub-Miocene Surface; Early Caenozoic Surface), 110, 113–116, 118, 124, 125
Allophane, 83
Alluvia, 11
Alteration of detritus, 35, 36, 78, 96, *Plate 11*
Alteration of high-level laterite (*see* Induration; Karst; Mesas, postincision modifications)
Alteration of mechanical residuum (*see also* Re-solution of mechanical residuum), 23, 36, 92, 106, 107
Alteration of remanié (*see* Alteration of mechanical residuum)
Alternating conditions of wetting and drying (*see also* Iron, segregation; Water table oscillation)
  and laterite formation, 21, 40, 42, 43, 45, 54

Altitude of laterite-capped mesas (*see* Mesas, altitudinal distribution)
Alumina (*see also* Bauxite), 3, 12, 49, 70, 82, 83, 85, 88, 89
  accumulation of, 50, 90
  as a definitive characteristic of laterite, 13–18
  influence of lithology on alumina content of crusts, 25–28
  mobilization of, 4, 83, 93
Aluminium (*see* Alumina)
Aluminosilicates (*see* Silicates of alumina)
Aluminous laterite (*see* Bauxite)
Aluminous nodules, 88
Aluminous pisoliths, 70
Aluminous rocks (*see individual rock types*), 25
Alumogels, 83
Amorphous aluminium, 82
Amorphous iron (*see* Colloidal substances; Organic complexes; Ferric hydroxide sol; *see also* Limonite), 80
Amorphous silica, 90
Amphibolites, 24, 29, 30
Anatase, 86
Ankole, 111, 115
"Apparent peneplain" (*see* Mesas, acyclic development)
Artefacts, 21, 48
Atmospheric climate, 41–45
Australia, 1, 41, 58. 60

B

Bacteria, 4
  precipitation of iron and manganese by, 84
  reduction of ferric iron by, 81
Banding (*see also* "Cutane"; "Patina"), 69 70
Basal sapping (*see also* Mesas, postincision modifications), 61

141

Copper, 89
Cretaceous, 109, 113, 125
Cryptocrystalline aggregates of aluminium, 83
Crystalline basement complex rocks, 27
Crystalloid substances, 69
Cuba, 52, 87, 88, 107
Cuirasse, 16
Currently forming laterite, 21
Cutane, 70, *Plates 2, 9, 11*
Cycle of erosion (*see* Erosion cycle)
Cyclic significance of laterite (*see also* Planation surface laterite; Mature laterite; Immature laterite; Acyclic development of laterite), 22, 30, 32–34, 37, 38, 78, 79

D

Dating of planation surfaces
    by fossil evidence, 110, 112, 114, 115
    by morphological continuity, 110, 111, 113
    relative, 115–123
Datum, 105, 125
Dead laterite, 21, 22, 31
Deccan, 15
Deep weathering (*see* Weathering)
Deforestation (*see also* Burning; Vegetation deterioration), 43, 53, 61, 99, 106, *Plate 6*
Deformation of laterite surfaces (*see* Mesas, postincision modifications)
Deformation of planation surfaces (*see* Tilting; Warping; Faulting)
Dehydration of laterite (*see also* Exposure to air), 13, 50
Denudation chronology, 105, 108–125
Deposition of iron (*see* Iron precipitation)
Derived laterite (*see* Detrital laterite)
Desilicification, 7, 13–19, 25, 30, 40, 52, 83
Detrital laterite, 3, 34–36, 38, 56, 65, 67, 70, 73, 74, 78, 95–97, 105, 107, 116, 117, 121, 122, *Fig. 32, Plates 2, 11, 17*
Detrital models of laterite formation, 94–97, 107
Detrital pisoliths
    diagnostic characteristics, 70, *Plates 2, 9, 11*
Detritus (*see* Detrital laterite; *see also* Mechanical residuum)
Diagnostic characteristics of African Surface,
    altitude, 116
    flatness, 110

inselbergs, 110
    laterite, 116, 117
Diaspore, 82
Differential postincision settling (*see also* Mesas, postincision modifications), 122
Differential solution, 2–5, 51, 59
Discrete concretion structures (*see* Oolite; Ooliths; Nodules; Shot; Pellets; Mottles; Pisoliths; Pisolite; Concretions; Microconcretions; *see also* Immature laterite) 37, 66–74, 77
Dissolving of laterite under trees (*see* Solution, of laterite under trees)
Dolomite, 27
Downward leaching of iron, 49–52, 54, 55, 57–61, 65, 78, 81, 82, 85, 86, 94, 95, 99, 102, 106, 107
Downward movement of iron (*see* Downward leaching of iron; Mechanical residuum; Detritus)
Down-wasting, 8, 51, 52, 71, 72, 92, 96, 101–105, 107, 119, *Plate 10*
Drainage conditions (*see also* Waterlogging; Water table)
    and bauxite formation, 83
    and kaolin crystallinity, 90
    and laterite structures, 27, 37
    and lateritization, 32
Drainage patterns in Uganda, 111, 117, 123, 124
Duricrusts, v, 16, 41, 93, 116

E

Early Caenozoic Surface (*see also* African Surface; Mid-Tertiary Surface; Miocene Surfaces; Sub-Miocene Surface), 110
Earth movements (*see* Isostacy; Tilting; Warping; Faulting; Rejuvenation; Folding), 109, 110, 125
East Africa, 73, 109, 114, 116
Eastern Africa, 48
Edaphic climax grassland, 46–49, 52, *Plates 4, 6, 7, 15*
Enclosed hollows (*see also* Mesas, post-incision modifications), 50, 61, 122
End-Tertiary Surface, 110, 113, 114, 116
Enrichment of laterite (*see* Source of enrichment; *see also* Mechanisms for iron translocation)
Entebbe, 111, 114
Entebbe graben, 114
Epidiorites, 24

143